About IFPRI

The International Food Policy Research Institute (IFPRI), established in 1975, provides evidence-based policy solutions to sustainably end hunger and malnutrition and reduce poverty. The Institute conducts research, communicates results, optimizes partnerships, and builds capacity to ensure sustainable food production, promote healthy food systems, improve markets and trade, transform agriculture, build resilience, and strengthen institutions and governance. Gender is considered in all of the Institute's work. IFPRI collaborates with partners around the world, including development implementers, public institutions, the private sector, and farmers' organizations, to ensure that local, national, regional, and global food policies are based on evidence.

IFPRI is a member of the CGIAR Consortium.

2012 GLOBAL FOOD POLICY REPORT

IFPRI

INTERNATIONAL FOOD POLICY RESEARCH INSTITUTE

International Food Policy Research Institute
2033 K Street, NW
Washington, DC 20006-1002, USA
Telephone: +1-202-862-5600
www.ifpri.org

ISBN: 978-0-89629-553-7

DOI: 10.2499/9780896295537

Cataloging-in-Publication Data is available from the Library of Congress.

PHOTO CREDITS

Cover image: Bangkok, Thailand, © 2007 Jeff Hutchens/ Getty Images
Chapter images: page x, G20 Summit at Los Cabos, Mexico, © 2012 Andres Leighton/Associated Press; page 14, Limeira, Brazil, © 2012 Paulo Whitaker/Reuters; page 28, Bauko, Philippines, © 2010 G. M. B. Akash/Panos; page 38, Harjbook Ki Bhagal, Rajasthan, India, © 2011 Dieter Telemans/Panos; page 48, Kamwosor Centre Village, Kenya, © 2011 Russell Powell, Courtesy of Heifer International; page 58, Fremont, Nebraska, United States, © 2012 Nati Harnik/Associated Press; page 68, Inner Mongolia, China, Georg Gerster/Panos; page 88, Chiang Mai, Thailand, Jean-Leo Dugast/Panos.

Cover design: Carolyn Hallowell / Book design and layout: Carolyn Hallowell, Lucy McCoy, and David Popham.

Contents

FEATURES

Preface

This *2012 Global Food Policy Report* is the second in an annual series that provides an in-depth look at major food policy developments and events. Initiated in response to resurgent interest in food security, the series offers a yearly overview of the food policy developments that have contributed to or hindered progress in food and nutrition security. It reviews what happened in food policy and why, examines key challenges and opportunities, shares new evidence and knowledge, and highlights emerging issues.

In 2012, world food security remained vulnerable. While talk about hunger and malnutrition was plentiful, it remains to be seen whether current and past commitments to invest in agriculture, food security, and nutrition will be met. New data from the Food and Agriculture Organization of the United Nations suggest that the world will fall short of achieving the first Millennium Development Goal of halving the prevalence of undernutrition by 2015. Translating commitments into action is thus even more urgent.

Evidence points to a number of steps that would advance food and nutrition security. Investments designed to raise agricultural productivity—especially investments in research and innovation—would address one important factor in food security. Research is also needed to investigate the emerging nexus among agriculture, nutrition, and health on the one hand, and food, water, and energy on the other. In addition, by optimizing the use of resources, innovation can contribute to the push for a sustainable "green economy." Boosting agricultural growth and turning farming into a modern and forward-looking occupation can help give a future to large young rural populations in developing countries.

Beyond investing in research and innovation, more can be done in other areas to improve food security. Donors can pay more attention to gender when designing development projects. Policymakers can take into account the global repercussions of their domestic agricultural policies. This is true not only for the developed countries, but also for Brazil, China, and India—emerging players in world agricultural markets whose actions will have a sizable impact on future food security. Building poor people's resilience to shocks and stressors would help ensure food security in a changing world. In any case, poor and hungry people must be at the center of the post-2015 development agenda.

The topics covered in the *2012 Global Food Policy Report* were selected following a number of consultations designed to capture the depth, relevance, and breadth of food policy issues in 2012. For inclusion in the report, a topic must represent a new development in or a new way of looking at a food policy issue; it must be international in scope (it must have affected several countries or stakeholders); and high-quality research results or expert judgment must be available to allow for authoritative discussion of the topic. To add perspectives and deepen discussion, we supplemented chapters with shorter contributions from experts and stakeholders, including farmer representatives from Africa, Asia, and Latin America. I would like to thank IFPRI's Board of Trustees and Strategic Advisory Council, as well as IFPRI staff, for their insights on current food policy developments and their expert advice on the selection of key issues.

I hope that this publication is met with interest, informs stakeholders and decisionmakers the world over, helps set the research agenda for 2013 and beyond, and contributes to improving food policies so they benefit the world's poorest and most vulnerable people. I welcome your feedback, comments, and suggestions at ifpri@cgiar.org.

SHENGGEN FAN
Director General

Acknowledgments

The *2012 Global Food Policy Report* was prepared under the overall leadership of Shenggen Fan and a core team comprising Rajul Pandya-Lorch, Gwendolyn Stansbury, Alexander J. Stein, and Klaus von Grebmer. The report benefited greatly from the strategic insights of a committee of advisers and other experts, including Robert Bos, Betty Bugusu, Margaret Catley-Carlson, Marion Guillou, Craig Hanson, Michiel A. Keyzer, Will Martin, Kimberly Pfeifer, Martin Pineiro, Prabhu Pingali, Florence Rolle, M. S. Swaminathan, Eric Tollens, Joachim von Braun, and Emorn Wasantwisut.

Excellent text and data contributions were made by Luz Marina Alvaré, Kwadwo Asenso-Okyere, Suresh Babu, Ousmane Badiane, Nienke Beintema, Samuel Benin, Prapti Bhandary, Clemens Breisinger, Karen Brooks, Jean-Christophe Bureau, Jean C. Buzby, Kevin Z. Chen, Dyborn Chibonga, Morgane Danielou, Nitin Desai, Eugenio Díaz-Bonilla, Paul Dorosh, Keith Fuglie, Amy Gautam, Caren Grown, Peter Hazell, Paul Heisey, Emily Hogue, P. K. Joshi, Hansdeep Khaira, Sergey Kiselev, David Laborde, Sylvie Lemmet, Tsitsi Makombe, Ruth Meinzen-Dick, Nancy Morgan, Siwa Msangi, André Meloni Nassar, Gerald Nelson, Alejandro Nin-Pratt, Jemimah Njuki, Tolulope Olofinbiyi, David Orden, Ma. Estrella A. Penunia, Adam Prakash, Agnes Quisumbing, Muhammad Abdur Razzaque, Claudia Ringler, Mark W. Rosegrant, Gert-Jan Stads, Gwendolyn Stansbury, Alexander J. Stein, Simla Tokgoz, Jeremías Vasquez, Klaus von Grebmer, Elizabeth Waithanji, Bingxin Yu, Funing Zhong, and Sergiy Zorya.

Production of the report was led by IFPRI's Publications Unit under the guidance of Andrea Pedolsky; Heidi Fritschel provided superb writing support, and the overall production of a high-quality report was made possible through the dedicated work of Adrienne Chu, Carolyn Hallowell, Pat Fowlkes, Corinne Garber, Michael Go, Lucy McCoy, David Popham, Ashley St. Thomas, Julia Vivalo, and John Whitehead. In addition, Chapter 1, which draws partially on other chapters in this book, benefited from valuable research and writing assistance from consultant Joanna Brzeska.

Importantly, the report also benefited from the thorough work by IFPRI's Publications Review Committee, chaired by Gershon Feder, as well as from the anonymous scholars and experts who peer-reviewed the research and provided insightful comments on the preliminary drafts.

G2012
LOS CABOS
MÉXICO
SAUDI ARABIA
OECD
FAO
OECD

Walk the Talk

Shenggen Fan

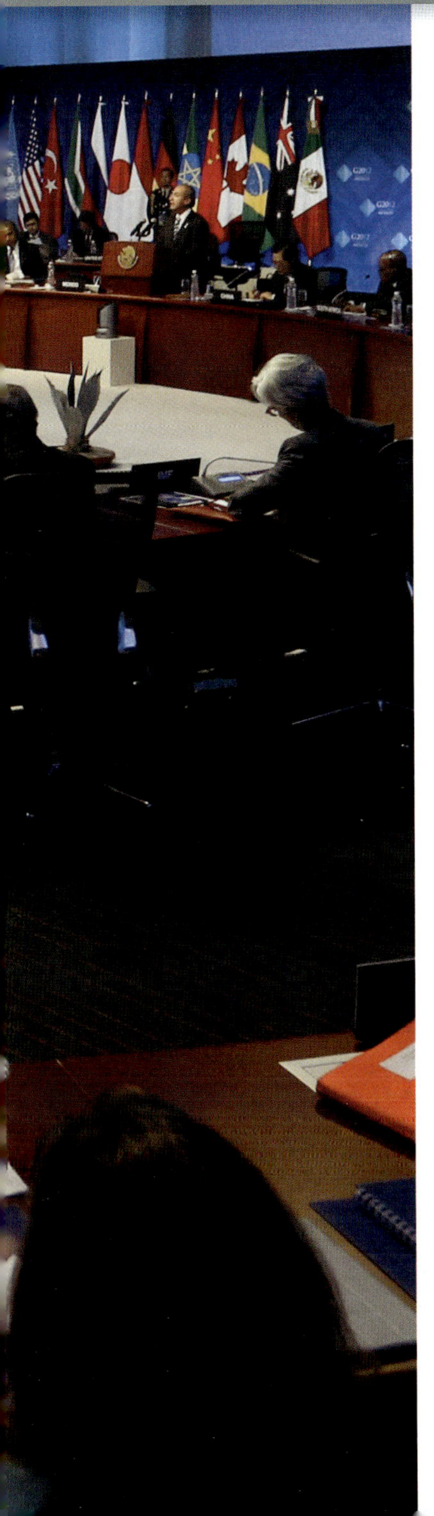

THE WORLD FOOD SYSTEM CONTINUED TO BE IN A VULNERABLE position in 2012. As the 2015 deadline for the Millennium Development Goals approaches, progress toward halving the proportion of people suffering from hunger is not on track. Granted, a number of countries made important and promising changes in food-related policies, and the global community made noteworthy commitments to strengthen aspects of food security. For 2013, however, it will be critical to ensure that the discussions and commitments made in previous years are translated into concrete actions.

Talk about food security in global and regional bodies was abundant in 2012. It was widely agreed at the Rio+20 summit that incorporating environmental sustainability into economic policies and activities offers opportunities for achieving "green growth." Two goals—Zero Net Land Degradation and the Zero Hunger Challenge—were launched during the event. Much was said during the Group of 20 (G20) and Group of Eight (G8) summits about the need to increase investment in agriculture, especially in research to enhance agricultural productivity and food security, and investment in nutrition to enhance long-term human capital. But there is a need for a mechanism to ensure and monitor actual implementation. New commitments were also made to calm global food price volatility and spikes and to increase transparency in bulk land acquisition deals through the adoption of voluntary guidelines for land investment.

In a landscape full of rhetoric and promises, the on-the-ground reality of implementation and action was mixed. On the one hand, there were several positive developments, often built on the strong base established in recent years. A number of countries in Africa have made noteworthy progress transforming agriculture into a more productive and sustainable sector. Agricultural spending, including investments in agricultural research by emerging economics such as Brazil, China, and India, continued to increase. Donor support to international agricultural research, particularly to CGIAR, maintained momentum after the strong growth of 2011. The private sector further

---

Shenggen Fan is the director general of the International Food Policy Research Institute in Washington, DC.

enhanced its commitment to global food security through active engagement in the Business 20 (B20) summit and with the G8. Development agencies such as the United States Agency for International Development (USAID), the United Kingdom Department for International Development (DFID), the German Federal Ministry for Economic Cooperation and Development (BMZ), the International Fund for Agricultural Development (IFAD), and the World Bank scaled up their investments in agriculture, food security, and nutrition.[1] Private foundations such as the Bill & Melinda Gates Foundation also continued to play a larger role in international agricultural development in 2012.

On the other hand, progress fell short of previous commitments. Only a handful of African countries met their 10 percent target of agricultural spending as a share of the national budget. Emerging economies and some African countries often used increased agricultural spending to subsidize inputs and outputs, leading to trade distortions; overuse of fertilizer, water, and energy; and the crowding out of productive investments in areas such as agricultural research and development (R&D), irrigation, and rural infrastructure. To protect domestic consumers, several countries continued to use trade export bans, which exacerbate global food price volatility. Global trade in food and agriculture remained protected, and the prospects of reaching any more trade agreements appear remote. The United States and European Union proposed new agricultural policies that could potentially distort world agricultural and food markets, leading to adverse effects on smallholder farmers in developing countries. International negotiations on climate change in December were disappointing, and the agreement made in Cancun, Mexico, in 2011 to decide on an agricultural work program was ignored, with discussion deferred to June 2013.

THE GLOBAL FOOD SYSTEM REMAINED FRAGILE IN 2012

New Numbers, Same Problem

Despite the revision of the methodology and data used by the Food and Agriculture Organization of the United Nations (FAO), the new numbers of hungry people paint only a slightly more optimistic picture of the undernourished population (Figure 1). Older estimates had shown an upward trajectory for world hunger and a spike in 2008 and 2009. The new estimates, however, show that developing countries made significant and constant progress in reducing chronic undernourishment until 2007—when progress slowed—and are closer than previously believed to reaching the Millennium Development Goal of halving the prevalence of undernourishment by 2015. Still, the number of chronically undernourished people remained high, at 870 million in 2010–2012.

Although the new methodology and data are important steps toward obtaining a more comprehensive measure of food security, these numbers also leave much information out. They reflect the quantity of food in people's diets (specifically, kilocalories), but not the quality (that is, vitamins and minerals). It is believed that a large number of people suffer from micronutrient deficiencies ("hidden hunger"), which is not captured by the new measure.

Drought and Volatile Food Prices

The 2012 droughts in Central Asia, Eastern Europe, and the United States led to tighter cereal supplies and, subsequently, a spike in world cereal prices. The year 2012 was the hottest year on record in the United States.[2] Approximately 80 percent of farmland in the United States was hit by the most severe drought in half a century, with maize and soybeans the most affected.[3] Similarly, high temperatures and low rainfall reduced wheat production in Australia, Kazakhstan, Russia, and Ukraine, which are among the top producers and exporters of wheat. Because the most affected regions are also some of the largest producers of key staple crops, the drought will have implications for global food

FIGURE 1 Estimates and projections of undernourished people worldwide, 1990–2015

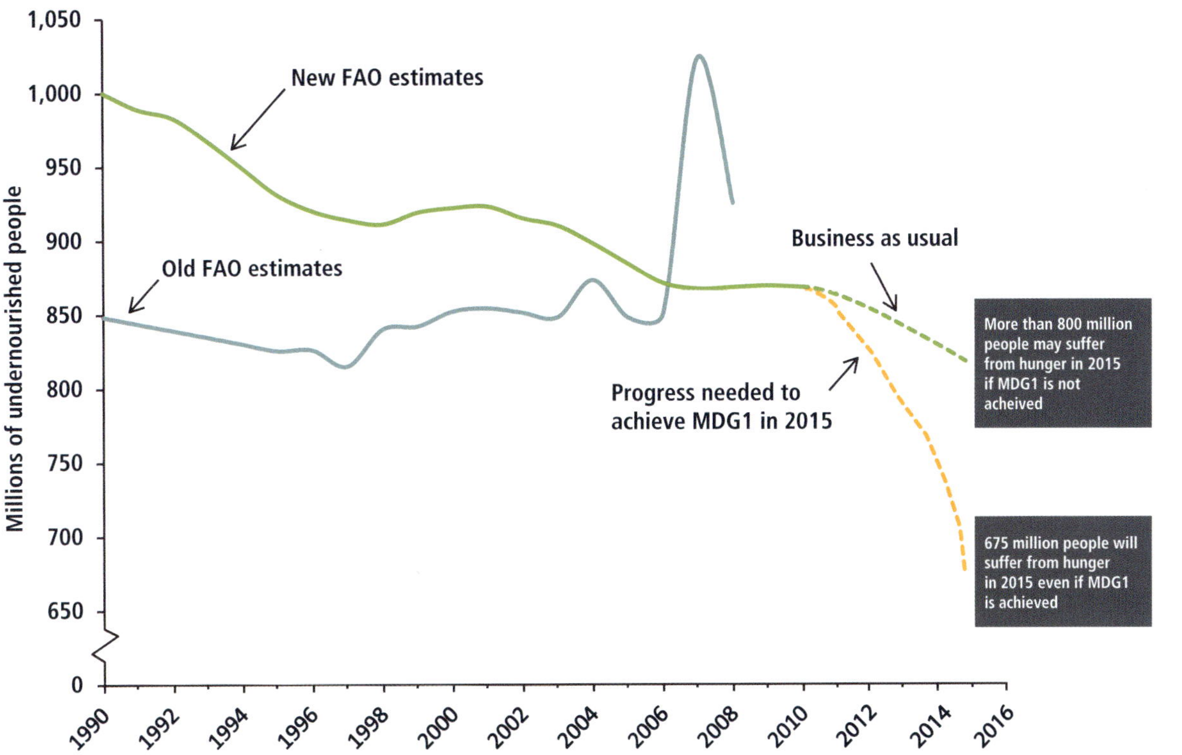

Sources: Old estimates are from Food and Agriculture Organization of the United Nations (FAO), *The State of Food Insecurity in the World* (Rome, various years); new estimates are from FAO, *The State of Food Insecurity in the World 2012* (Rome, 2012); author's projections are based on data from FAO and the United Nations.

security well into 2013 through upward pressure on food prices. Erratic rains and prolonged dry spells throughout Southern Africa also resulted in declining maize production in Lesotho, South Africa, and southern parts of Malawi, Mozambique, and Zimbabwe. Crop production rebounded in the Sahel region in Africa following the 2011 drought, but closer inspection of the Sahel crisis suggests that current food insecurity there is more a reflection of the region's chronic, long-term vulnerability than the result of a sudden, short-term shock such as a single drought in 2011.[4] The resilience of chronically vulnerable communities in the region to crisis is weak. People barely had time to recover and rebuild already limited assets after previous droughts before the 2011 drought hit the region.

Drought conditions, together with the continued use of maize for biofuel, contributed to a 25 percent increase in international maize prices between June and August, with prices reaching record levels in August.[5] Because maize is used not only for human consumption, but also for livestock feed, higher maize prices led to higher prices for animal-based products, and this increase is predicted to continue in 2013. Similarly, the international price of wheat rose by 32 percent between June and August 2012, although it was still well below 2008 levels. The increase in international cereal prices has proven to be especially problematic for countries that depend on cereal imports, in particular in Central and South America and Central Asia.

Prices for many coarse grains also soared in the Sahel region in 2011 and 2012 owing to a combination of drought, civil unrest, and locust infestation. In Burkina Faso and Mali, millet prices rose by 66 and 63 percent, respectively, compared with 2011; sorghum prices, by 52 and 43 percent; and maize prices, by 21 and 44 percent.[6] Toward the

2012 FOOD POLICY
ISSUES, ACTIONS & DECISIONS

TOWARD RESILIENCE IN THE SAHEL
Stakeholders create a Global Alliance for Resilience Initiative to help West African nations better cope with future food crises.
June 18

G8 COMMITS TO FOOD AND NUTRITION SECURITY IN AFRICA
G8 members, African countries, and private-sector leaders support the New Alliance for Food Security and Nutrition.
May 19

NIGERIA SETS AMBITIOUS 2030 GOAL
Working with the private sector, the Ministry of Agriculture mobilizes to create an agricultural sector worth $256 billion.
June 7

CHINA PRIORITIZES INVESTMENT IN AGRICULTURAL SCIENCE
The government's No. 1 Document for 2012 chooses accelerating agricultural science and technological innovation as its theme.
February 1

VOLUNTARY GUIDELINES FOR LAND TENURE ADOPTED
The Committee on World Food Security endorses voluntary guidelines for safeguarding the rights of people to own or access land, forests, and fisheries.
May 11

O AGREES TO BOOST RICULTURAL ODUCTIVITY

e G20 agrees to promote eater public and private estment in agriculture d technology.

ne 19

STAINABLE DEVELOPMENT IS FT WANTING AT RIO+20

e UN's Rio+20 Declaration offers ong vision but little direction on v to achieve food security in a een economy.

ne 22

DROP IN US RAINFALL, PEAK IN GLOBAL FOOD PRICES

The worst drought in the United States since the 1950s severely lowers its maize and soybean production and drives up prices on world markets.

August 22

US FARM BILL EXPIRES

Congress recesses until after the November elections without passing a new farm bill, leaving the agricultural sector up in the air.

September 30

NEW WAY TO CRUNCH THE GLOBAL HUNGER NUMBERS

The Food and Agriculture Organization of the United Nations publishes lower estimate of the number of undernourished people—which remains unacceptably high.

October 9

NEW LIMITS ON FOOD CROP–BASED BIOFUEL PRODUCTION IN THE EU

To stimulate development of alternative biofuels from nonfood feedstock, the EU proposes to limit global land conversion for food crop–based biofuel production.

October 17

NO WHEAT FROM UKRAINE

Government limits grain exports informally, destabilizing markets.

November 15

UN CONFERENCE EXTENDS KYOTO PROTOCOL TO 2020

Many believe the results of the UN's 18th conference on climate change are inadequate to contain global warming at 2 degrees Celsius.

November 26

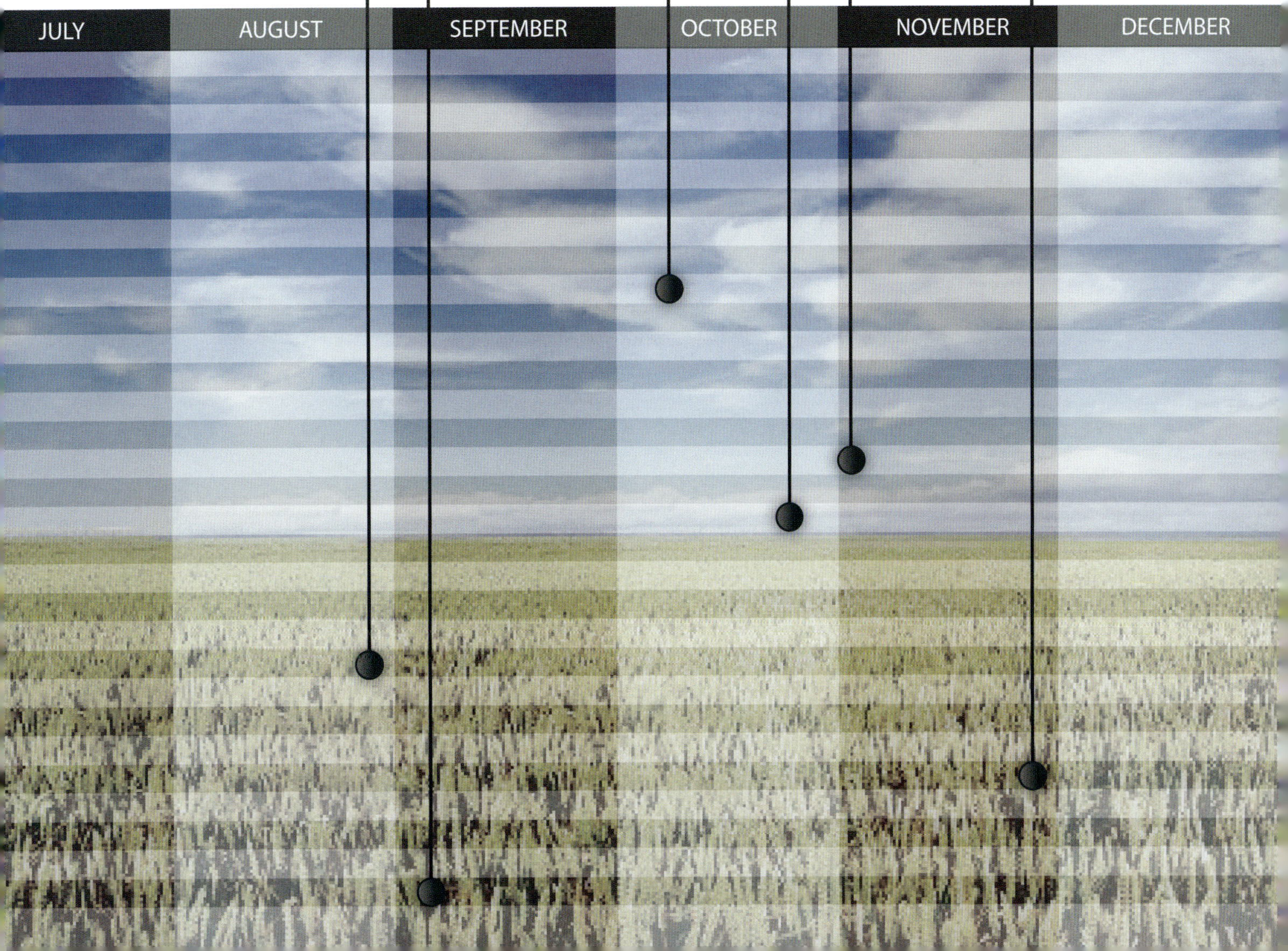

end of 2012, rising cereal supplies—as a result of increased domestic production and cross-border imports—contributed to a decline in prices from record or near-record levels. Nonetheless, the decline has been small and prices continue to be relatively high because of flooding and political insecurity in the region, threatening the food security of nearly 19 million people.[7] Similarly, the increase in wheat and maize prices in Southern Africa in 2012 due to low rainfall and international market pressure has contributed to the deterioration in the food security of approximately 1.5 million more people in the region since 2011.

In response to rising grain prices and decreasing production and stocks, a number of countries—including Argentina, Malawi, and Zambia—imposed or maintained restrictions or bans on grain exports. Such restrictions are designed to protect domestic food security during times of reduced food production and higher food prices, but in reality they are often counterproductive and result in market instability and price volatility. International wheat prices increased slightly amid fears that Russia and Ukraine would impose bans on grain exports to protect domestic sellers and buyers, but they soon fell again when both countries ultimately rejected the bans. Some still fear that these two countries will use informal mechanisms, such as increasing administrative barriers and limiting access to infrastructure, to limit exports. In Tanzania, research has shown that maize export bans have had a detrimental effect on the rural poor and agricultural growth,[8] and in response to this evidence, the Tanzanian government lifted its ban in autumn 2012.

Conflict

Violent conflict—both a cause and a consequence of food insecurity—played a role in a number of countries in 2012. Food security in Central Africa is especially hampered by persistent conflicts that send thousands of internally displaced people and refugees flooding into already strained food systems. An armed conflict in northern Mali, followed by a military coup in the capital, led to the displacement of more than 400,000 people (internally and in neighboring countries), disrupting trade flows, putting pressure on already limited local food resources, and worsening the precarious and drought-ridden food security situation in neighboring countries created by the poor harvest in 2011.[9] Renewed violence in the Democratic Republic of Congo resulted in the displacement of approximately 2.2 million people within the country and forced a further 70,000 people to flee to neighboring countries.[10] Furthermore, the Democratic Republic of Congo is also home to more than 100,000 refugees from other conflict-ridden countries in the subregion. Ongoing conflict and low rainfall have also disrupted food security in Somalia, particularly in the southern and central parts of the country.[11] All of these conflicts have halted agricultural and livestock production activities, markets, and trade in the affected countries, depriving many poor households of their livelihood and limiting food access and availability.

In the Arab region, civil war in Syria has led to a refugee crisis that is being compounded by a food crisis. In late 2012, the World Food Programme identified about 1.5 million Syrians as being in urgent need of food assistance.[12] Elsewhere in the region, in the aftermath of the Arab Awakening, political and economic recovery has been slow. Food security in Tunisia has stabilized, but because of lower crop production and foreign currency inflows during 2012, the food security situation has worsened in Yemen and to a lesser degree in Egypt and Libya.

Long-Term Drivers of the Global Food System

A number of strong driving forces are exerting pressure on food production, consumption, and markets. For example, rising incomes and rapid urbanization in many developing countries are changing the composition of food demand. Energy markets are having a greater impact on food security thanks to growing biofuel markets and the increasing share of energy in agricultural costs. The recent push for more investment in agricultural R&D may result in advances in agricultural productivity that also have a large impact on food systems.

The implications of these changes for food security will depend heavily on the choices policymakers make now and in the years ahead. IFPRI's International Model for Policy Analysis of Agricultural Commodities and Trade (IMPACT) is a useful way to look at future scenarios under different policies and other conditions. Researchers have used it to look specifically at what could happen to food prices and food security if the world achieves higher agricultural productivity, experiences higher energy prices, or lowers demand for meat. The results show that policy choices that lead to higher energy prices could make food prices even higher and more volatile than they have been in recent years. Cutting developed countries' consumption of livestock products has only small impacts on food security in developing countries. Increasing agricultural productivity, however, by expanding public and private investments would lower food prices, lead to higher agricultural production, and result in greater food security. Changes in the dietary patterns, productivity growth, and energy policies of emerging countries—namely Brazil, China, and India—have an especially large impact on future food security outcomes given the significant roles of these countries as producers and consumers (see Chapter 8 of this report).

DEVELOPMENTS AND OPPORTUNITIES IN 2012

New Sources of Agricultural Growth

After years of stagnation, new evidence in 2012 showed that in many developing countries the transformation of agriculture into a modern, competitive, and productive sector accelerated in recent years. Between 2001 and 2010, world agricultural production grew at an average annual rate of 2.4 percent, close to its historical average growth rate of 2.3 percent a year since the 1970s. Closer inspection reveals that agricultural production entered a period of accelerated growth around 1995, following more than 20 years of gradually decreasing growth rates. At the same time, there has been a global shift in what kind of food is grown and where. Global food production increasingly comes from developing countries and is made up of less cereal grains and more horticultural and oil crops, a reflection of changes in the types of foods consumers are demanding.

The sources of agricultural growth have changed over the past several decades. Growth in total factor productivity, a measure of output growth that does not come from input growth, accelerated substantially in 2001–2009 compared with the average for 1971–2009 (see Chapter 2). Until the late 1980s, farmers achieved most of the growth by using more inputs such as land, fertilizer, and labor. This input intensification accounted for 90 percent of agricultural growth in the 1960s, 80 percent in the 1970s, and 75 percent in the 1980s. Starting in the 1990s, however, greater use of inputs accounted for less than 20 percent of agricultural growth, while more than 80 percent came from higher total factor productivity—that is, producing more with the same amount of inputs.

Two large developing countries in particular, Brazil and China, have had sustained high growth in total factor productivity over the past two decades. Several other developing regions, including Southeast Asia, West Asia and North Africa, and Latin America and the Caribbean, also registered accelerated total factor productivity growth during the last decade. The major exception is Africa south of the Sahara, where long-run total factor productivity growth has remained below 1 percent a year. Rapid growth in Africa still comes largely from farmers' cultivation of new land and greater use of fertilizers. This situation points to the great potential for Africa to accelerate its growth and transform its agricultural sector through an increase in agricultural productivity.

Investments in agriculture—especially in agricultural research and innovation—have been shown to play an especially important role in raising agricultural productivity, overcoming constraints posed by increasingly scarce resources such as land and water, and improving economic efficiency in the use of fertilizers and pesticides.

Pushing to a Green Economy

The "green economy" movement took a prominent place on the international stage at the United Nations Conference on Sustainable Development in Rio de Janeiro in June. At this event, known as

Rio+20, heads of state were challenged to commit to a concept of a green economy that integrates sustainable development and poverty eradication (see Chapter 3). The resulting declaration, "The Future We Want," offers a vision of sustainable development and calls for a wide range of actions, including the integration of ecological concerns into overall economic policy and the formulation of sustainable development goals.[13] Despite the vision presented, the conference did not produce a firm policy roadmap and timeframe. The indicators of commitment and measures of accountability that are needed to realize the vision are also clearly lacking.

During the Rio+20 event, the United Nations secretary-general launched the Zero Hunger Challenge in an ambitious bid to combine hunger reduction with sustainable development efforts. The initiative calls for access to adequate food all year round for all people, the elimination of stunting in children younger than two, the sustainability of all food systems, a 100 percent increase in smallholder productivity and income, and zero loss and waste of food. Similarly, leaders agreed to targets of zero net land degradation by 2030, zero net forest degradation by 2030, and drought preparedness policies in all drought-prone countries by 2020. While these initiatives should be applauded, they need to be accompanied by clear measures, timeframes, and accountability mechanisms to become a reality.

One aspect of the green economy that has increasingly been featured in policy discourse and research is the idea of the bioeconomy—an economy that has moved beyond petroleum and is based on the use of renewable bio-based resources to produce food, health, and industrial products and energy. The bioeconomy includes not only crop agriculture, but also aquaculture, forestry, and biomass production, among many other things. In early 2012, the European Commission presented its bioeconomy strategy and action plan with the objective of increasing investments in bioeconomy research and enhancing the competitiveness of bioeconomy sectors.[14]

Gender: From Attention to Action

The year 2012 was marked by significant new attention to the role of gender equality in agricultural growth and food security. A wealth of recent evidence has shown that agricultural and nonagricultural reforms to increase women's capacities, engagement, and access to productive resources can improve agricultural performance and food security (see Chapter 4). The World Bank's *World Development Report 2012* and FAO's *State of Food and Agriculture 2010–2011,* for example, emphasized women's important contributions to agriculture in developing countries, highlighting the agricultural productivity gains and nutritional benefits that can be reaped from greater gender equality. Increasing food security requires policies that most efficiently close the gender gap in women's access to resources and services within and outside of agriculture, including education, extension, technologies, political institutions, and financial services.[15]

In response to the evidence, aid donors and developing countries have taken steps in recent years to address gender inequality and the special needs of women in agriculture as part of broad-based food security programs. Until recently, however, measuring the degree to which development programs actually empower women has been difficult. In 2012, to help quantify changes in women's empowerment and gender equality, IFPRI, USAID, and the Oxford Poverty and Human Development Initiative launched the Women's Empowerment in Agriculture Index.[16] The index—a tool for monitoring how agricultural development programs affect women's empowerment and gender equality—is currently used to assess programs under Feed the Future, an initiative led by USAID and executed by various US government agencies in a number of developing countries.

The renewed commitment to gender equality in agriculture can result in improved productivity and food security, but it must first be mainstreamed into policy actions. Agricultural strategies and programs need to be based on a deeper understanding of the similar and different interests of women and men as both consumers and producers within food systems, paying specific attention to the gender gap in access to assets. To achieve this, greater efforts are needed to collect evidence disaggregated by gender that can be used to improve future interventions and keep all actors accountable.

Where the Jobs Are

Unemployment and underemployment have significant social and economic implications. Agricultural development in many developing countries has the potential both to improve food security and to create jobs. However, young people in many developing countries often do not see farming as a viable and lucrative career, and they reject agriculture in favor of jobs in cities. Yet, in order for agriculture to become a technically dynamic and high-productivity sector that contributes to food security, it needs an influx of educated and innovative young labor (see Chapter 5). In turn, a profitable and productive agricultural sector would provide employment opportunities for a growing population of young people and thereby raise both their food production and their incomes. The engagement of young people in the agricultural sector is especially important in Africa, where the manufacturing and services sectors in urban areas cannot fully absorb the burgeoning young population entering the labor force. A growing and diversifying agricultural sector will also fuel the development of the rural nonfarm economy, especially the services sector, which can also play an important role in generating income and employment opportunities for young people by providing the agricultural sector with vital services and goods.

In 2012, several international meetings signaled a greater focus on the goal of increasing young people's employment in agriculture. The "Young People, Farming, and Food" conference, held in Ghana, examined research and policies related to engaging young people in the agrifood sector. The 4th Conference of the African Union Ministers in Charge of Youth, held in Ethiopia, highlighted the need for countries to implement the African Union's strategies for empowering youth and increasing youth-focused investments. The International Fund for Agricultural Development's Farmers' Forum included a special session on youth in agriculture, which proposed increasing youth representation in farmers' organizations, increasing funding for agricultural programs that target youth, and improving young people's access to natural resources, markets, financial services, and knowledge.[17] Some country-level efforts to support young commercial farmers have also been made. For example, the government of Nigeria has placed a significant emphasis on agriculture in its recently launched Youth Employment Programme and taken significant steps toward launching a Youth Employment in Agriculture Programme, including a consultative workshop, with the goal of creating 1 million jobs for youth by 2015.

Youth employment in agriculture and the nonfarm economy can take various forms, including full-time work on existing or new landholdings, small-scale agribusiness such as veterinary services or mechanization services, retail trade, or wage labor on farms or in agricultural processing plants. To support these options, developing countries must make agricultural land, capital, and skills more available to young people and make agricultural development programs more responsive to their needs. Countries must promote innovations in rural financial services and institutions, deregulate land rental markets, provide demand-driven agricultural advisory services, and offer flexible short-term training programs. Additionally, young people in rural areas need to have easy access to the amenities that are available in urban areas, including physical and social infrastructure such as roads, electricity, and education. Agriculture in Africa, for example, needs to be seen not only as an instrument for economic growth and improved food security but also as a major employer of the region's young people. Making agriculture profitable, competitive, and dynamic will not only attract young people, but also benefit the wider society and global community by increasing growth, improving food security, and preserving an increasingly fragile natural environment. To support these efforts, developing countries need to link their existing political commitments to agricultural development and youth employment in order to capture the complementarities of the agricultural and youth agendas.

Extending Support for Rich-Country Farmers

Despite years of calls for an end to high and distortionary payments to farmers in Europe and the United States, no such change occurred in 2012 (see Chapter 6). A new US farm bill was not passed in 2012; rather, the US Congress extended the previous farm bill for one year, meaning that the issues raised in 2012 will continue to be debated in 2013. The proposed new bill actually increases support for agriculture by replacing the current system of annual fixed payments to farmers with insurance subsidies designed to protect farmers from both annual and multiyear losses. Given the apparent evolution of US farm policy toward greater domestic support for agriculture, this bill could make it more difficult for the United States to build global support for freer trade and less distortionary agricultural policies.[18]

The European Commission (EC) has proposed a number of key reforms to its Common Agricultural Policy. Over the past 20 years the Common Agricultural Policy has progressively shifted away from price supports and subsidized exports of surpluses toward direct payments to farmers. The current proposal makes the payments more conditional on farmers' compliance with environmental regulations, requiring farmers to, among other things, diversify crops, maintain existing permanent pastures, and dedicate a minimum amount of land to ecological focus areas. Income transfers to farmers will remain very high (compared with the United States). The proposal also includes measures to help farmers cope with future price crises, albeit to a much lesser degree than in the United States, given that the EC proposes ceilings on any insurance and income stabilization payments to farmers.

Although these support policies are not directly linked to production incentives, they can help farmers maintain or increase their production levels by improving their production capacity. If the support policies raise production in the European Union and the United States, this could potentially distort and undermine agricultural production in other countries and contribute to a risky concentration of agricultural production in just a few countries. These policies could launch the global food system on a slippery slope, where other countries adopt their own distortionary agricultural policies.

At the same time, however, a positive step took place with regard to biofuel policy in Europe. In response to the growing debate over the use of crops for food versus fuel, the EC introduced a proposal in October 2012 to impose a 5 percent limit on the use of food crop–based biofuels to meet the European Union's 10 percent renewable energy target by 2020. This proposal aims to promote the development of alternative second- and third-generation biofuels, which contribute significantly less to greenhouse gas emissions and do not compete with global food production. In the United States, despite numerous discussions and calls for change in 2012, particularly after the recent drought, little was done to reduce the use of food grains in biofuel production.

Food Policy Developments across Regions and Countries: A Mixed Picture

A number of significant food policy developments took place in all major regions in 2012. Although these changes often received less attention than global initiatives or events, they nonetheless have fundamental impacts on global food security (see Chapter 7).

African agriculture continued along its path of transformation in 2012. From 2006 to 2011, annual agricultural growth was strong in a number of African countries: approximately 12–13 percent in Angola and Liberia; 7 percent in Botswana, Ethiopia, and Malawi; 5 percent in Rwanda; and 4 percent in Ghana and Tanzania.[19] This rapid growth was fueled largely by more investment in agriculture, increased use of fertilizer, and the adoption of high-yielding crop varieties, together with more friendly macroeconomic policies. Poverty rates in these countries have declined, but rates of hunger and malnutrition remain high. And the region is extremely vulnerable to weather shocks and conflict. The 2011–2012 food crisis in the African Sahel region was a testimony to this vulnerability. In response to the crisis, the Global Alliance for Resilience Initiative in the Sahel was launched to protect vulnerable communities through a combination of social safety nets, nutrition programs, emergency food reserves, and assistance with raising people's production and assets. In the long run, however, making the region more resilient and reducing its vulnerability will require raising agricultural productivity through policy reforms and investments in agricultural research, irrigation, market infrastructure, and institutions.

In South Asia, India's decision to allow foreign direct investment in multibrand retail—such as supermarkets—in 2012 may provide an opportunity to transform the country's private retail sector through increased competition and investments. By potentially cutting out intermediaries and improving storage technologies and transportation, this reform could reduce food prices for poor consumers while offering small farmers better and more profitable market access, thus improving India's food security. However, because the increase in foreign direct investment could also have negative impacts, such as putting small vendors out of business, emphasis needs to be put on the net effects of the reform. Along the same lines, Bangladesh's food security has improved significantly over the past several years because of the government's deliberate policies to increase investment in agriculture, scale up social safety nets, and build national grain stocks for emergency and social safety net purposes. The government has also developed an ambitious vision to substantially reduce hunger and poverty by 2021 and an investment plan to secure the country's food supply. The country was expected to produce record levels of food in 2012, but localized food insecurity persists.

At the same time, there have been some reversals in policy developments in the region. India's Technical Expert Committee recommended a 10-year moratorium on field trials of genetically modified crops, depriving the country of the chance to use these technologies to further enhance yields, nutrition, and resilience against droughts, floods, heat, biotic stresses, and other natural adversaries. The Supreme Court has not yet made a final decision on the moratorium and is currently awaiting a more comprehensive report on genetically modified crops from the committee.

In Nepal, agriculture can play a key role in improving food and nutrition security, but the government's proposal to substantially increase fertilizer subsidies may do more harm than good by crowding out more productive investments—in areas such as R&D, irrigation, and rural infrastructure—and social safety programs. This is especially problematic and significant in Nepal, which has one of the highest rates of child malnutrition in the world.

In East Asia, the year 2012 was a turning point for China. After many years of near self-sufficiency in major grains (rice, wheat, and maize), China had to import 2–3 million metric tons of rice, 4–5 million metric tons of wheat, and 5–6 million metric tons of maize in 2012, in addition to continuing to import almost 60 million metric tons of soybeans. This shift may lead the country to pursue protective trade policies in the future, with potentially negative implications for famers in other developing countries and for the global agricultural trade system. A positive development was China's 2012 No. 1 Document, which laid out the government's plan for substantial investment in agricultural R&D in 2012 and beyond. The country's public spending on agriculture is outpacing that of Brazil and India and is on a trajectory approaching that of the United States.

The traditional leader in rice exports—Thailand—exported less rice than India and Vietnam in 2012. Thailand's new policy of guaranteeing farmers' prices at levels well above the market rate pushed up the price of its rice and made it uncompetitive, leading to a reduction in its rice exports and an increase in government rice stocks. Although the policy pushed international rice prices slightly higher, the impact was largely neutralized by strong competition and increased rice exports from India on the heels of the Indian government's removal of restrictions on certain types of rice exports in 2011.[20] Myanmar took steps toward increasing its rice exports by undertaking agricultural reforms, creating an agricultural bank, and supporting seed production companies. If the right policies are adopted, the country has the potential to become a major food exporter in the region.

Latin America and the Caribbean produced an increasing share of the world's agricultural output in 2012 compared with previous years. Although extreme weather may have reduced 2012 grain production, meat production was projected to increase. Public agricultural investment has increased in Latin America in recent years, but most of this increase has taken place in just a few countries. Investment in smaller, poorer countries has declined. It is also worth noting that Latin America has a dual role to play both as a large exporter of agricultural and food products and as a provider of environmental goods, such as rainforests with rich biodiversity. In 2012, intense debate between Brazilian farmers and environmentalists over the use of the Amazon rainforest continued. Brazil has successfully developed a rapidly growing and robust sugarcane-based ethanol industry over the past several decades, and now the government and private sector in Brazil are working to replicate this success with the development of biodiesel production from palm oil. This expansion will undoubtedly mean that biofuel production will continue to have an important impact on biodiversity and global food security.

The Arab world is one of the few regions where hunger levels have increased in recent years. Economic growth in the region has been persistently slow, and child malnutrition is high. To address food insecurity, governments have mostly continued the policies adopted during the 2008 global food crisis and the 2010–2011 uprisings, such as increased public sector wages and subsidies for fuel and food—policies that are often poorly targeted and that strain already stretched public budgets. Yet some Arab countries also initiated longer-term policies and investments in 2012 designed to improve food security, such as reforming food subsidies (Jordan, Sudan, and Tunisia), increasing grain reserves (the Gulf Cooperation Council[21]), and setting up a committee to monitor food prices and availability (Saudi Arabia).

What to Watch for in 2013

- Where will agriculture, food security, and nutrition be positioned in the post-2015 development agenda, and how will they be accounted for if the green economy moves from concept to reality?
- How effective will the international development community be in strengthening the resilience of the global food system to shocks and emergencies? Will we see improvements in the early-warning and food security information systems in developing countries?
- To what extent will climate change be taken more seriously at intergovernmental forums? Will binding commitments come forth?
- How will economic uncertainties in the United States and the European Union influence the quantity and quality of their foreign assistance policies? And what will be the impact of their new agricultural policies on food security in developing countries?
- Will China continue to increase its grain imports, or will there be additional institutional innovations to transform its agricultural sector?
- Will India's food security bill and large-scale programs such as direct cash transfers improve its food and nutrition security?
- Can the continued conflicts in the Arab World and Africa be contained, or will further unrest shake the regions and affect food security?
- How will the private sector, in particular multinational food companies, engage in addressing malnutrition—both undernutrition and obesity—in developing countries?
- With major changes on the environmentalist front regarding the use of genetically modified organisms in agriculture, will there be a significant shift in Europe's acceptance levels with potential positive outcomes for global food security?
- How will increased information sharing—through the open access movement and the soaring use of mobile devices—affect agriculture and rural development?

OUTLOOK FOR 2013

Many of the factors that have caused today's vulnerable food security will remain in 2013. Poor countries and poor people will continue to be hard hit by a number of ongoing economic and environmental shocks, such as natural disasters, conflicts, and the lingering volatility of food prices. Given the likelihood of such shocks, there is an urgent need to build the resilience of global and national food systems as well as of poor households and people. This means developing strategies and policies that help individuals, communities, regions, and countries cope with and recover from shocks and achieve food security, health, and well-being. Social systems and ecosystems should be made resilient to both natural disasters and human-induced crises, both sudden and slow moving. Dry areas deserve special attention as they are home to half of the world's poor and hungry people, and climate change will make them even more vulnerable. International dialogues, such as the World Economic Forum, the G8, and the G20, must be used as platforms to develop this concept, propose policy options, and formulate concrete commitments and actions to reduce poor people's vulnerability to food and nutrition insecurity and enhance their capacity for long-term growth.

The nexus among agriculture, nutrition, and health, and the nexus among food, water, and energy have been promoted separately for the past several years, including at the "Bonn 2011 Conference: The Water, Energy, and Food Security Nexus" and IFPRI's 2011 conference "Leveraging Agriculture for Improving Nutrition and Health." The agenda for 2013 must bring all of these themes together. Although trade-offs exist among agriculture, nutrition and health, and environmental sustainability, it is important to explore and develop triple-win solutions. Measurable goals should thus include both environmental sustainability and nutrition in addition to agricultural growth and food security and should have clear timelines and phases. Efforts to protect and promote human health should go hand in hand with efforts to improve the health of livestock, crops, and ecosystems. To build up evidence on triple-win mechanisms, indicators must be developed to track and evaluate the food security implications of green economy strategies, policies, and investments. Many conferences on nutrition will take place in 2013 and 2014, such as the Joint FAO/World Health Organization International Conference on Nutrition. These discussions should take an integrated approach to tackling the challenge of malnutrition, including establishing targeted safety nets, reshaping agricultural policies and practices to promote nutrition, reducing food waste and losses, promoting sustainable and healthy diets, and increasing the coverage of clean drinking water and sanitation.

Greater technical and financial support should continue to be allocated toward establishing national institutions to design, implement, monitor, and evaluate food security programs, initiatives, and policies. Asian experiences have shown that returns from this type of support are high, and many countries have committed resources and initiated policies to support food security and poverty reduction. The greatest challenge they face is the lack of capacity in implementing these programs, initiatives, and policies. The G8, which is under the leadership of the United Kingdom in 2013, must fulfill the commitments made in L'Aquila, Italy, in 2009 and support the implementation of country and regional agricultural strategies and plans through country-led coordination processes. In a welcome development, Ireland has put hunger high on the agenda of its European Union presidency during the first half of 2013.

> Much was said during the G20 and G8 summits about the need to increase investment in agriculture and food security. But there is a need for a mechanism to ensure and monitor actual implementation.

Finally, in 2013 and beyond, the conversation will continue on the Millennium Development Goals and their successors. Past progress, while inadequate in many ways, has shown that cutting food insecurity—sometimes dramatically—is possible. There will also be a push to integrate environmental sustainability goals into the post-2015 development agenda. While development goals should strive for environmental sustainability, poor people must be the center of the post-2015 development agenda. The focus of food policies should shift from cutting hunger toward eliminating it completely—within a clear timeframe and with mechanisms for holding countries, international institutions, and other relevant actors accountable for meeting this goal.

The global development community has been busy talking about overcoming hunger and malnutrition for long enough. Now it is time to walk the talk by turning the discussions and promises into actions. ■

Chapter 2

AGRICULTURAL PRODUCTIVITY

A Changing Global Harvest

Keith Fuglie and Alejandro Nin-Pratt

In 1961 the world was feeding 3.5 billion people by cultivating 1.37 billion hectares of land. A half century later, the world population had doubled to 7 billion while land under cultivation increased by only 12 percent to 1.53 billion hectares. How, then, did agricultural production triple? By increasing productivity. By getting more output from existing resources, global agriculture has grown, proving wrong past concerns that the world's population would exceed its food supply. In fact, at the global level, the long-run trend since at least 1900 has been one of increasing food abundance: in inflation-adjusted dollars, food prices fell by an average of 1 percent per year over the course of the 20th century (Figure 1). But then, over the past decade, something changed.

Around 2002, real food prices began to rise, and the shock was not merely a temporary one. Agricultural commodity prices spiked sharply in 2008, 2010, and again in 2012. Demand-side factors (including continued population growth, greater per capita consumption of meat, and diversion of crop commodities for biofuel) and weather-induced production shocks (like the 2012 drought in North America) are certainly major forces behind the high and volatile prices of recent years. But the persistence of rising commodity prices has renewed concerns about whether agriculture is facing new constraints on growth. In fact, for major cereal grains like wheat and rice, average rates of yield growth have slowed from about 2 percent per year in the 1970s and 1980s to about 1 percent per year since 1990. Additionally, there is evidence that some developed countries have recently seen a slowing down of growth in agricultural total factor productivity (a broad measure of sectorwide productivity), which has an effect on developing and developed countries alike.[1] A slowdown in agricultural productivity growth could signal rising food scarcity, higher commodity prices, and increased competition for the world's land, water, and energy resources. With such grave consequences, it is more urgent than ever to

---

Keith Fuglie is the chief of the Resource, Environmental, and Science Policy (RESP) Branch at the US Department of Agriculture's Economic Research Service (USDA-ERS) in Washington, DC. **Alejandro Nin-Pratt** is a research fellow in the Development Strategy and Governance Division of the International Food Policy Research Institute in Washington, DC. The views expressed in this chapter are those of the authors and do not necessarily reflect those of the USDA-ERS.

Cutting Consumer Food Waste

JEAN C. BUZBY

Industrialized countries waste more food per capita than developing countries. For example, in 2007 North America and Europe wasted 95–115 kilograms of food per capita, compared with 6–11 kilograms per capita in Africa south of the Sahara and South and Southeast Asia.[1] Few peer-reviewed, published studies provide national food waste estimates, particularly for farm-level losses. Nevertheless, the food waste literature suggests that most of the food waste in industrialized countries occurs at the consumer level (not at the farm level, as in developing countries). Waste also represents lost resources used to produce that food.[2] This means that soil is eroded, water sources depleted, and air possibly polluted for food that never even gets consumed.[3]

It would, of course, be ideal to just generate less waste overall. As a supplemental strategy, the US Environmental Protection Agency's[4] "food recovery hierarchy" suggests that the top priority is to recover or claim wholesome food before it is wasted to feed hungry people by, for example, donating it to local food banks. Using food waste that meets safety standards for livestock, zoo animals, and pets is next in the hierarchy, followed by recycling food and food waste for industrial purposes. Composting food to improve soil fertility is a relatively low priority because the focus is to first make the most of the resource material before returning it to the soil. The last resort should be disposal through landfilling or incineration because of the negative impacts on the environment.

Food waste occurs for many reasons. Many of these causes are similar across industrialized countries (for example, food often spoils when consumers buy more than they need with family-sized packaging or "buy 1, get 1 free" offers), but some factors have greater variation and are less understood (such as food used in cultural traditions). Regardless, food waste at the consumer level is so widespread—occurring every day in millions of households, food-service venues, schools, hospitals, and other institutions worldwide—that interventions will be challenging. Diverting uneaten food to the next best use involves resource and logistical challenges, but perhaps the success story of recycling can provide helpful information.

Understanding where and how much food is wasted and the value of this waste is important information that industries and policymakers can use to raise awareness, reduce food waste, and increase the efficiency of both the farm-to-fork system and food recovery efforts to feed the growing population. Governments may be able to work with the food industry and consumer groups to motivate reductions in food waste at every stage of the food chain.

Jean C. Buzby is an economist in the Food Economics Division of the US Department of Agriculture's Economic Research Service in Washington, DC.

ensure agricultural productivity growth. This chapter offers a reassessment of that growth at the global level and identifies ways to keep it on the upswing.

WHAT CHANGED?

World agriculture has undergone some fundamental changes in the past few decades. One has been that many developing countries have greatly expanded their capacities in agricultural research and innovation. Combined with support from international agricultural research centers, this has led to the availability of improved technologies and practices for local farmers. Complementing this have been institutional and policy reforms, improvements in farmer education and health, and investments in rural infrastructure, all of which help create an environment where new farm technologies and practices are adopted more rapidly. Greater productivity growth in developing-country agriculture can certainly pull up the average for global productivity.

A second major development has been the changing location and composition of global agricultural production. With slower agricultural growth in developed countries and a significant

FIGURE 1 Agricultural price index and population trend, 1900–2010

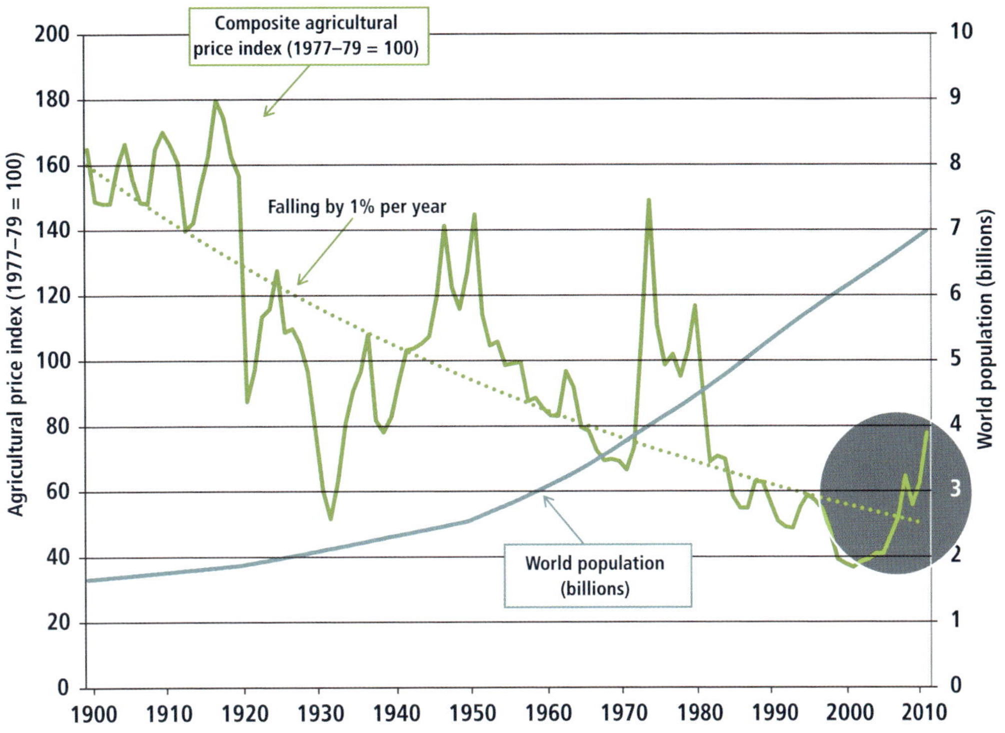

Source: K. O. Fuglie and S. L. Wang, "New Evidence Points to Robust but Uneven Productivity Growth in Global Agriculture," *Amber Waves* 10 (September 2012).

reduction in agricultural output from post-Soviet states, developing countries now account for a large and growing share of global agricultural production. And, as rising incomes cause changes in the types of foods consumers demand, the share of staple food commodities in world agricultural production has declined. Two new studies—one published in 2012 and one that is forthcoming—used different methods to estimate trends in agricultural productivity at the global level.[2] Both found that the productivity growth rate has actually accelerated in recent decades, led by improved performance in developing countries. It follows, therefore, that future challenges to global food security, apart from long-term risks related to climate change, are more likely to be the result of uneven access to resources, technologies, and food than the world's ability to increase global agricultural production and food availability in the aggregate.

A SHIFTING AGRICULTURAL SUPPLY

World agricultural production grew at an average annual rate of 2.4 percent between 2001 and 2010, close to its historical average growth rate since the 1970s of 2.3 percent per year. However, recent years demonstrate a period of accelerated growth that started around 1995, which, in turn, followed more than 20 years of gradually decreasing growth rates (Figure 2).

The exceptionally slow growth observed in the 1990s reflected a sharp contraction in agricultural production in the former Soviet bloc, but the trend of declining agricultural growth in the decades prior to 1995 also includes a slowing of growth in some high-income countries, especially in Western Europe and Japan.

This slowing of growth in high-income and transition economies of the former Soviet bloc has led to a major geographic shift in where agricultural production takes place globally (Figure 3). In 1965, 56 percent of total agricultural output was

FIGURE 2 Evolution of the annual growth rate of global agriculture, 1970–2010

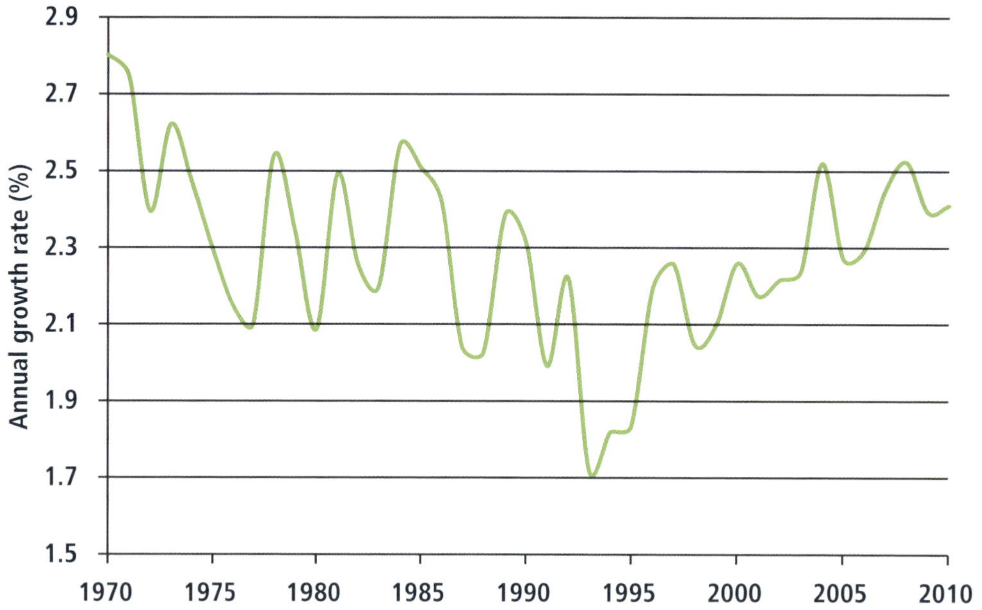

Source: Elaborated by authors using data from FAOSTAT, accessed May 2012.

produced in those same countries, although they only comprised 33 percent of the world's population at that time. Developing countries, on the other hand, with 76 percent of the world population, produced just 44 percent of total agricultural output. By 2010, the same high-income and transition economies produced 32 percent of global

> In addition to the shifting location of agricultural production, changes have occurred in its composition.

agricultural output and held 21 percent of world population. Developing countries accounted for 68 percent of global agricultural output, with East, Southeast, and South Asia contributing 44 percent (and comprising 52 percent of world population), and Latin America, Africa, and West Asia contributing the remaining 24 percent of global agricultural output (and comprising 27 percent of world population).

Within developing regions, Northeast Asia (dominated by China) has sustained agricultural growth rates averaging more than 4 percent per year since 1971 (Table 1). Southeast Asia, West Asia and North Africa, and Latin America and the Caribbean also achieved rapid growth in agricultural output, at around 3 percent per year, while agricultural growth in Africa south of the Sahara averaged significantly lower (2.4 percent per year).

In the 1980s, about half of the total growth in global agriculture came from East, Southeast, and South Asia, a contribution that reached 70 percent in the 1990s and 60 percent in the 2000s. High-income and transition countries contributed about 30 percent of the growth in global agricultural supply in the 1980s, but this fell to practically zero in the 1990s (with negative growth in the transition countries during this decade) before recovering to about a 10 percent share of global agricultural growth in the 2000s. The importance of Latin America and the Caribbean has increased over time, and, in the 2000s, the region accounted for nearly 17 percent of the growth in global agriculture.

FIGURE 3 Share of total agricultural production, by regions and groups of countries

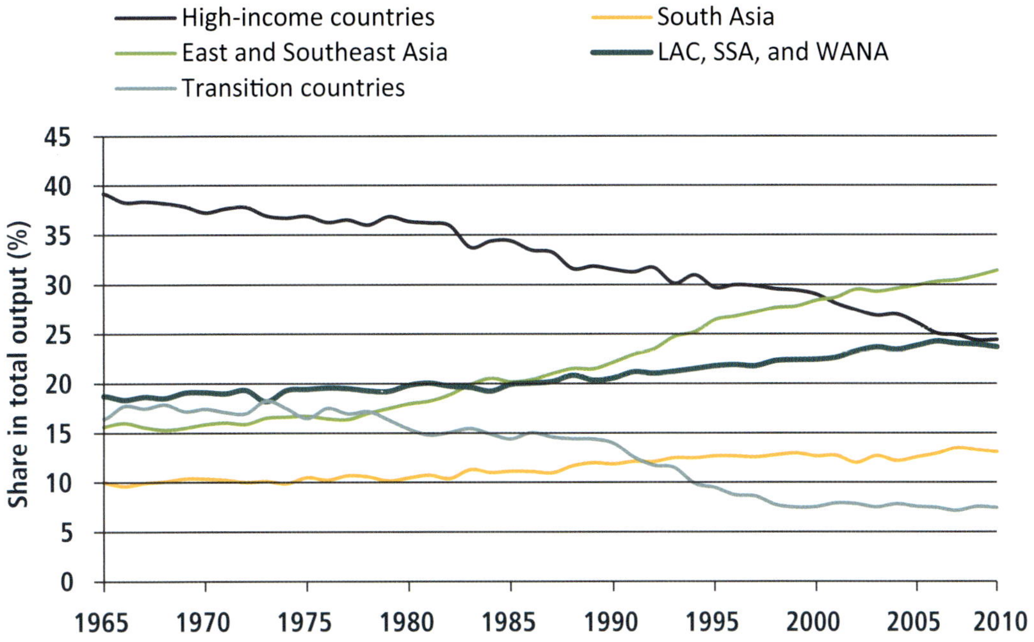

Source: Elaborated by authors using data from FAOSTAT, accessed May 2012.
Notes: LAC = Latin America and the Caribbean; SSA = Africa south of the Sahara; WANA = West Asia and North Africa.

In addition to the shifting location of agricultural production, changes have occurred in its composition (Figure 4). While the share of livestock products (meat, milk, eggs, hides, and wool) in total agricultural output has remained stable (around 37 percent from 1970 to 2009), the share of cereal grains has fallen significantly (from 25 to 21 percent of the total). Meanwhile, production of horticultural and oil crops has grown rapidly, with the share of total output from fruits and vegetables rising from 16 to 22 percent and oil crops from 6 to 8 percent over the same period. The changing composition of global agricultural output reflects changes in the types of foods consumers are demanding. With rising per capita incomes, especially in developing countries, demand is

TABLE 1 Average annual growth rates of agriculture, by region (%)

	1971–1980	1981–1990	1991–2000	2001–2010	1971–2010
High-income countries	1.83	0.97	1.25	0.47	1.14
Transition countries	0.81	1.42	-4.03	2.28	0.04
Developing countries					
LAC	2.93	2.35	3.09	3.21	2.89
Northeast Asia	3.23	5.04	5.04	3.39	4.19
South Asia	2.19	3.70	2.76	2.80	2.86
Southeast Asia	3.66	3.32	3.41	4.23	3.64
SSA	1.05	2.68	3.11	2.97	2.44
WANA	3.31	3.84	2.61	2.75	3.13
World	2.08	2.42	2.09	2.42	2.25

Source: Elaborated by authors using data from FAOSTAT, accessed May 2012.
Notes: LAC = Latin America and the Caribbean; SSA = Africa south of the Sahara; WANA = West Asia and North Africa.

FIGURE 4 Composition of total global agricultural output

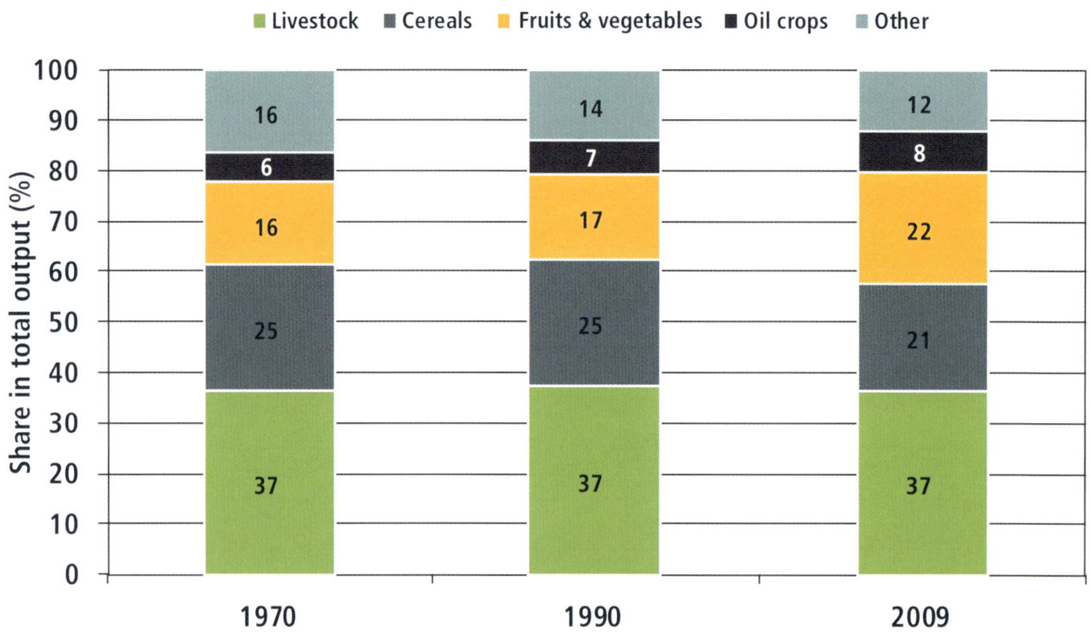

Source: Elaborated by authors using data from FAOSTAT, accessed May 2012.

shifting from staple food grains to more vegetables and fruits, vegetable oils, and animal products (and the protein-rich animal-feed meals provided by oilseeds as a co-product from crushing). Cereal grains, however, continue to supply 70–80 percent of the total caloric supply available for food, animal feed, and biofuel manufacturing.

The shifting location of world agricultural production to developing countries and the changing composition of agricultural output toward more horticultural and oil crops have significant implications for global trends in agricultural productivity. Increasingly, raising average global yields in crops and livestock relies on raising yields in developing countries. And moving production from relatively low-valued cereal crops to higher-valued horticultural crops can imply a rise in economic efficiency; by reallocating resources to produce commodities with greater value, farmers may improve productivity and income.

DRIVERS OF GROWTH: THE ROLE OF TOTAL FACTOR PRODUCTIVITY

Total factor productivity (TFP) measures the ratio of total commodity output (the sum of all crop and livestock products) to total inputs used in production, including all land, labor, capital, and materials. If total output is growing faster than total inputs, this is an improvement in TFP. An increase in TFP implies that more output is being produced from a given bundle of agricultural resources. TFP does not, however, take into account effects on environmental resources from agricultural activities, such as losses to biodiversity, nutrient runoff into water bodies, and greenhouse gas emissions.

Empirically, growth in TFP is generally measured as the difference in growth between outputs and inputs. Methods for measuring TFP differ mainly in the way in which outputs and inputs are aggregated. Figure 5 provides two estimates of long-run TFP growth in the global agricultural economy. One method uses a growth accounting approach ("TFP-growth accounting") in which

Reducing Postharvest Losses

NANCY MORGAN, ADAM PRAKASH, AND HANSDEEP KHAIRA

As global efforts are underway to ensure adequate and sustainably produced food for more than 9 billion people by 2050, the issue of postharvest losses has come to the forefront of the policy arena. These losses can occur for any number of reasons, including crop damage, spillage during transport, and biodeterioration during storage. Investing in ways to reduce these losses is a triple-win that would mean (1) improved food security, (2) greater food availability that alleviates pressure on prices, and (3) conserved valuable land, water, and labor resources.

Postharvest losses are clearly widespread, but quantifying total amounts is challenging; estimates—some as high as 50 percent—vary drastically from product to product, from system to system, and at different points along the supply chain. Similarly, the identification of what caused a loss—for instance, poor harvesting, inadequate storage, insufficient remuneration, or poor transport—is critical to determining the appropriate entry points for interventions.

The African Postharvest Losses Information System indicates that grain losses prior to processing in Africa south of the Sahara average between 10 and 20 percent. These losses are highly significant: if extrapolated for 2005–2007, they amount to nearly US$4 billion per year out of the estimated US$27 billion averaged overall production value.[1] This is on par with the US$3–7 billion in cereal that Africa imported annually between 2000 and 2007. If these losses were recuperated, they would allow 48 million people to consume the minimum 2,500 calories per day for a year. Similarly, the Food and Agriculture Organization of the United Nations estimates that approximately 1.3 billion tons of food are lost or wasted each year worldwide. In developing countries, per capita losses mainly occur at the production-to-retail nexus at around 120 kilograms per person in South and Southeast Asia and 200 kilograms per person in Latin America.[2]

By better understanding the magnitude of consequences brought about by postharvest losses along the food chain, we can leverage policies to improve food security, alleviate poverty, and sustain the environment. Filling in the data gap should be strategically complemented by interventions that range from using hermetically sealed bags and metallic silos to organizing producer associations that coordinate suppliers along the value chain. While these technologies and practices have proved useful, adoption rates in developing countries remain low. Identifying why requires an evaluation of failures and successes in the field, and an inclusive community of governments, practitioners, and donors can make that happen by sharing lessons and good practices. We need a revitalized approach for economically appropriate and socially relevant postharvest innovations that can be scaled up and used to inform national investment programs.

Nancy Morgan is a senior economist at the World Bank in Washington, DC. **Adam Prakash** and **Hansdeep Khaira** are statisticians at the Food and Agriculture Organization of the United Nations in Rome.

inputs are aggregated based on their share of total costs in production.[3] The second method uses a Malmquist productivity index estimated using data envelopment analysis ("TFP-DEA"), which aggregates inputs weighted by their opportunity cost or relative scarcity instead of using the actual market prices of inputs.[4] For comparative purposes, Figure 5 also shows long-run trends in the growth rates of agricultural output and land and labor productivity. Land and labor productivity typically show higher rates of growth than TFP because part of the growth in output per worker or output per hectare comes from more intensive use of other inputs, like capital or fertilizers, while TFP nets out the growth in these other inputs.

Both measures of TFP indicate that the average growth rate in global agricultural TFP accelerated between 1971 and 2009 (the latest year for which estimates are available), rising from less than 1 percent per year in the 1970s (according to both studies) to about 1.8 percent annually in 2001–2009 (using the TFP-growth accounting method) and 2.3 percent (using the TFP-DEA method).[5] Improvements in land productivity (total output

FIGURE 5 Productivity growth rates for global agriculture estimated using partial and total factor productivity measures

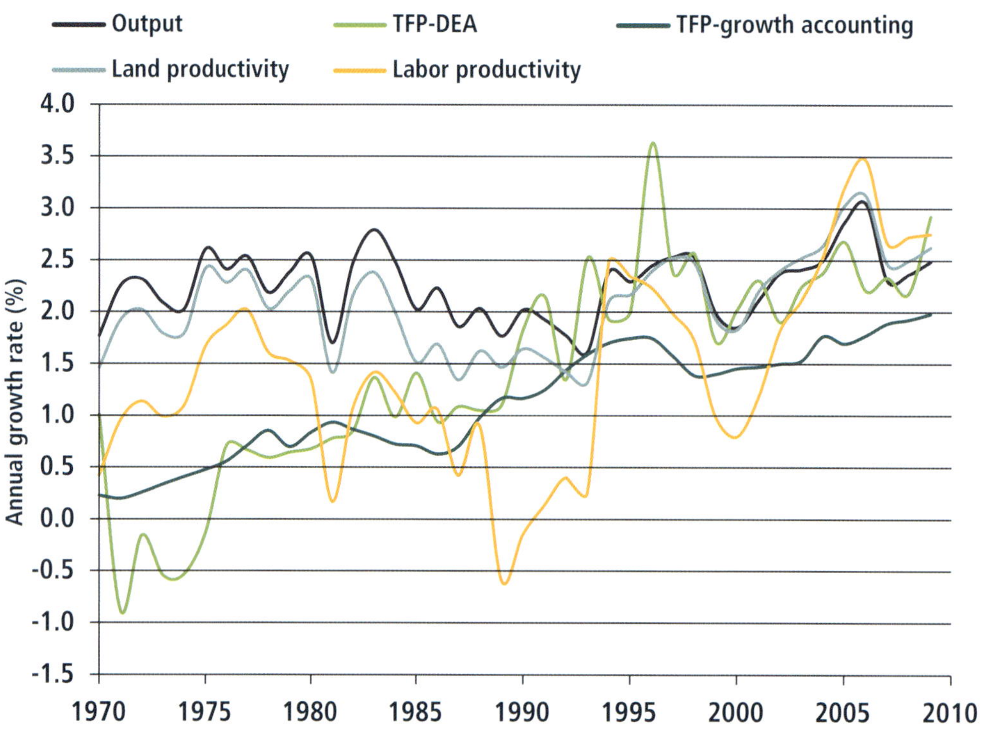

Source: Estimated by authors.
Notes: TFP-DEA rates are obtained using a Malmquist index and data envelopment analysis approach. TFP-growth accounting is estimated by aggregating inputs based on estimates of their cost share in production.

per hectare of agricultural land) have remained fairly steady at about 2 percent per year during the past 40 years while growth in labor productivity has also improved, but more slowly, reaching a rate of more than 2 percent per year only since the 1990s.

How much of the growth in output is due to increased resources, and how much of it is due to improved productivity? After nearly four decades of primarily resource-driven growth, a dramatic shift to productivity-driven increases in global agricultural output began around the early 1990s (Figure 6). Between 1961 and 2009, total resources and inputs grew about 60 percent as fast as growth in total agricultural output, implying that improvement in TFP accounted for only 40 percent of total output growth. But TFP's contribution to output rose over time, and between 2001 and 2009 it accounted for about 75 percent of the growth in global agricultural production. The contribution of natural resources (including land and water) to output growth has decreased gradually over time while that of input intensification (including the amount of labor, capital, and materials per hectare of land) has fallen sharply.

WHERE AGRICULTURAL PRODUCTIVITY IS GROWING AND WHY

Annual growth-rate estimates for land, labor, and total factor productivity are disaggregated among global regions in Table 2. Although the trends are hardly uniform, three general patterns are evident.

1. In high-income countries, the total amount of resources used in agriculture has been falling since about 1980. TFP growth offset the declining resource base to keep output from falling. TFP growth has remained robust overall but has

FIGURE 6 Sources of growth in global agricultural production

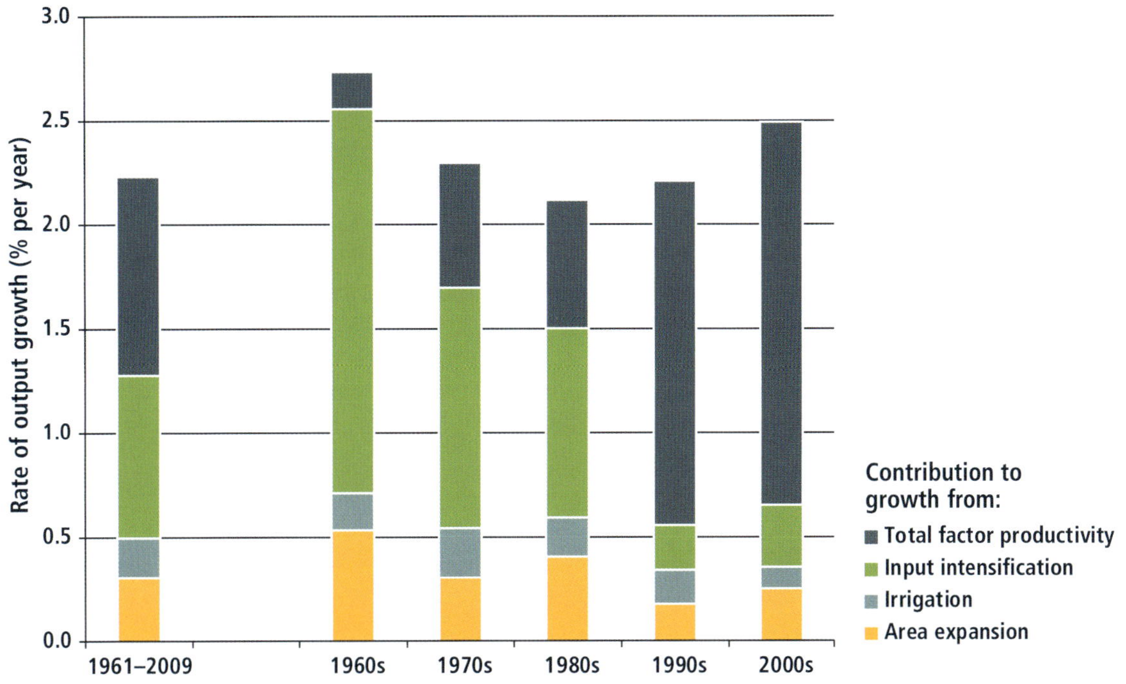

Source: K. Fuglie, "Productivity Growth and Technology Capital in the Global Agricultural Economy," in *Productivity Growth in Agriculture: An International Perspective*, ed. K. Fuglie, S. L. Wang, and V. Eldon Ball (Oxfordshire, England: CAB International, 2012).

slowed in some countries such as Australia and the United Kingdom. Labor productivity has been rising much faster than land productivity as the agricultural labor force in these countries declined and average farm size increased.

2. In developing regions, TFP growth saw substantial acceleration in 2001–2009 compared with 1971–2009. China and Brazil have sustained high TFP growth during the past two decades, and Southeast Asia, West Asia and North Africa, and Latin America and the Caribbean also

TABLE 2 Annual growth rates for land, labor, and total factor productivity, by region (%)

	2001–2009			1971–2009		
Region	Land	Labor	TFP	Land	Labor	TFP
World	2.37	1.94	1.22	2.06	1.26	0.65
High income	0.97	3.93	1.14	1.44	4.25	1.36
Transition	2.41	4.58	1.15	0.14	1.15	-0.13
Developing	0.82	0.36	1.29	1.11	0.24	0.28
LAC	3.38	4.12	1.30	2.62	2.79	0.53
WANA	2.52	2.08	1.33	2.38	2.44	0.42
China & Northeast Asia	3.72	5.26	1.34	3.80	4.05	0.69
South Asia	2.69	1.34	0.85	2.82	1.34	0.20
Southeast Asia	3.76	4.00	1.43	2.74	2.24	0.45
SSA	2.34	0.77	0.85	2.27	0.62	0.50

Source: Elaborated by authors using data from FAOSTAT, accessed May 2012.
Notes: LAC = Latin America and the Caribbean; SSA = Africa south of the Sahara; WANA = West Asia and North Africa.

demonstrated accelerated TFP growth in the 2000s. Africa south of the Sahara is the major exception, with long-run TFP growth staying below 1 percent per year.

3. In transition countries, the dissolution of the Soviet Union in 1991 imparted a major shock to agriculture. As they began the transition from centrally planned to market-oriented economies, agricultural resources sharply contracted and output fell. Since about 2001, however, output has begun to expand again, and it appears to be led by improvements in productivity. TFP growth, which was practically nonexistent during the Soviet era, has taken off since 2001.

New research has measured agricultural TFP growth not only for most countries, but also for various states and provinces within large countries, namely for Australia, Brazil, China, Indonesia, and the United States.[6] This work shows that productivity is highly variable not only across regions and countries but within them as well (Figure 7). In China, TFP growth has been very strong in coastal

What Makes African Agriculture Grow?

PETER HAZELL

After several decades of disappointing performance, the agricultural sector in Africa south of the Sahara has started to grow more rapidly. Exactly why it has begun to grow, however, and at what pace are points of contention. Reported agricultural growth rates vary depending on the methods and data used and the countries and time periods being evaluated. But generally they show that when measured in constant prices, agricultural gross domestic product (GDP) grew by between 2 and 3 percent per year from 1950 through 1999. This rate is consistent with estimates of the growth rates in agricultural production.

Since the late 1990s, Africa's agricultural GDP growth rate has been estimated to have increased by anywhere from 3 to 12 percent per year. Why such a wide variation? The global commodity price boom and higher inflation in the 2000s (and the way analysts account for those changes) had a big impact on estimates of the underlying agricultural growth rate. During 2000–2010, Africa's agricultural GDP grew by 12 percent per year in actual prices, 3.6 percent per year in constant prices, and 7.7 percent per year using the real increase in agricultural prices (that is, actual prices deflated by a cost-of-living index).[1] This higher estimate is closer to the 6 percent growth in real agricultural GDP reported during a similar period.[2] The lower estimate of 3.6 percent is consistent with estimates of the growth in agricultural production.

An increase from 2–3 percent to 3–4 percent in the annual growth rate of real agricultural GDP is not to be discounted, however, especially given the long period of neglect in agricultural investment that preceded it. For Africa to slash poverty and become food secure, the New Partnership for Africa's Development has targeted a 6 percent annual growth rate, so the faster the growth, the better. But, what's driving this faster growth?

In the past, most agricultural growth in Africa came from greater land and labor use, but the productivity (or incremental gain in production per unit of input used) of these and other factors (for example, fertilizers and improved seeds) remained low or declined. This pattern has now changed, with several studies reporting that factor productivity growth began to emerge as a more important driver of agricultural growth after the mid-1980s. Many of these gains were brought about by more efficient use of key factors following policy reforms in the 1980s and 1990s, whereas gains from improved technologies remain modest. This presents a challenge for future agricultural growth since the policy reforms have now run their course, and the opportunities to bring new land into farming are more limited, especially in many populous countries. Future agricultural growth will increasingly depend on technological change, which will require greater investment in agricultural research and development, rural infrastructure, and education.

Peter Hazell is an independent researcher and former division director at the International Food Policy Research Institute.

FIGURE 7 Average growth rate in agricultural productivity since the mid-1990s

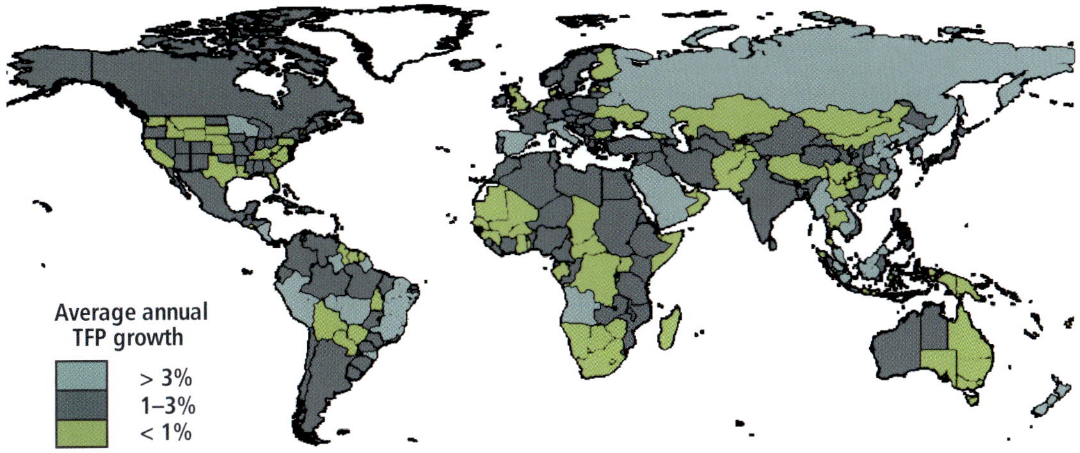

Source: K. Fuglie, "Productivity Growth and Technology Capital in the Global Agricultural Economy," in *Productivity Growth in Agriculture: An International Perspective*, ed. K. Fuglie, S. L. Wang, and V. Eldon Ball (Oxfordshire, England: CAB International, 2012).
Note: Growth rates are annual averages from 1995 to 2007 or 2009, depending on data availability.

provinces but less so in the interior. Coastal states of Brazil have also experienced robust agricultural productivity growth. But unlike China, high TFP growth is also evident in some parts of Brazil's interior—like Mato Grosso in the Cerrado, now the main soybean- and cotton-producing state in the country. In Indonesia, productivity growth has been concentrated in the western and northern regions of the country—Sumatra and Kalimantan, especially—where export commodities like oil palm have been booming. In contrast, TFP growth has been low or stagnant in Java and the eastern provinces. This is a sharp departure from the Green Revolution decades of the 1970s and 1980s, which disproportionately benefited irrigated rice production, an activity that is especially important in Java. In the United States, productivity growth has been moderately strong in agriculturally important areas such as the Corn Belt and the Great Lakes but low in the Great Plains, Appalachia, and major horticultural producers such as California and Florida. In Australia, broadacre (dryland) agricultural TFP has stagnated, primarily affecting the eastern and southern portions of the country.

Figure 7 also points to improved productivity growth performance in some African countries south of the Sahara since the mid-1990s. While a few raised their agricultural TFP growth to at least 1 percent per year, others (such as Angola) were simply recovering from earlier decades when they were at war. Africa south of the Sahara continues to face perhaps the biggest challenge in achieving sustained, long-term productivity growth in its agricultural sector. It is also the region of the world with the highest rates of poverty and food insecurity, and with the highest population growth rates projected for the coming decades.

Africa south of the Sahara continues to face perhaps the biggest challenge in achieving sustained, long-term productivity growth in its agricultural sector.

PROSPECTS AND FUTURE CHALLENGES

The development and adoption of improved farm technologies and practices has allowed food to become more abundant even as the population has grown and agricultural land has become scarcer. Advances in microbiology, information, and other

Agricultural R&D: Spending Speeds Up

NIENKE BEINTEMA, GERT-JAN STADS, KEITH FUGLIE, AND PAUL HEISEY

Systematic data on agricultural research and development (R&D) spending are greatly needed to identify areas where investment can lead to increased agricultural productivity and, ultimately, greater food security. IFPRI's Agricultural Science and Technology Indicators initiative collects this type of data and reported in its 2012 *Global Assessment of Agricultural R&D Spending* that between 2000 and 2008 (the latest year for which data are available) these R&D investments were on an upswing.[1]

Following a decade of slowing growth in the 1990s, global public spending on agricultural R&D increased steadily from $26.1 billion in 2000 to $31.7 billion in 2008.[2] Most of this growth was driven by developing countries while growth in high-income countries stalled; the increased spending in the former was largely driven by positive trends in a number of larger, more advanced middle-income countries (see figure in this box). China and India together accounted for close to half of the global increase of $5.6 billion. Other middle-income countries—particularly Argentina, Brazil, Iran, Nigeria, and Russia—also significantly increased their spending on public agricultural R&D during this period. These trends mask the negative developments that have taken place in numerous smaller, poorer, and more technologically challenged countries, which are often highly vulnerable to severe volatility in funding and subsequently see the continuity and viability of their research programs deteriorate. Many R&D agencies in these countries also lack the necessary human, operating, and infrastructural resources to successfully develop, adapt, and disseminate science-and-technology innovations.

Private investment in agricultural R&D also increased between 2000 and 2008—from $14.4 billion to $18.2 billion—and most of this R&D was carried out by companies in high-income countries. However, many of these companies have experiment stations in developing countries for the purpose of transferring new, proprietary technologies to those markets. Information on private-sector involvement in developing countries remains limited, but evidence suggests significant growth in large middle-income countries.

The combination of long-term sustainable government funding and a supportive policy environment has fueled increased agricultural productivity, as well as overall economic growth, in the world's more advanced developing countries, such as Brazil and China. Governments in the world's poorest countries need to make similar commitments or they will fall even farther behind.[3]

sciences are opening up new avenues for further improving agricultural productivity. As long as public and private investments in agricultural research and development are sufficient to translate these scientific advances into practical technologies for the many diverse farming environments and commodities worldwide, and as long as farmers can gain access to these technologies as well as markets for their produce, prospects seem bright for continued growth in global agricultural productivity.[7] Looking several decades ahead, the effects of a changing climate greatly increase uncertainties for agriculture and give further impetus to maintain and strengthen global capacities in agricultural science and technology.

Since the 1960s, long-term productivity growth in developing-country agriculture has been guided by three main pillars: (1) development of national capacities in agricultural research and innovation, (2) support from international public research centers and the private sector that provides better genetic materials and modern inputs, and (3) creation of an enabling environment for the rapid adoption of new technologies, including rural institutions that provide financial and educational services, rural infrastructure that improves access to markets, and economic and trade policies that allow markets to signal resource allocation.[8] Although productivity growth in developing-country agriculture remains uneven, many developing countries can still experience large leaps in productivity by using these pillars of growth, which were the foundation of the Green Revolution.

As in the past, achieving food security for all of the world's people requires more than raising agricultural productivity at the global level. Instead,

GLOBAL PUBLIC AGRICULTURAL R&D SPENDING, 2000–2008

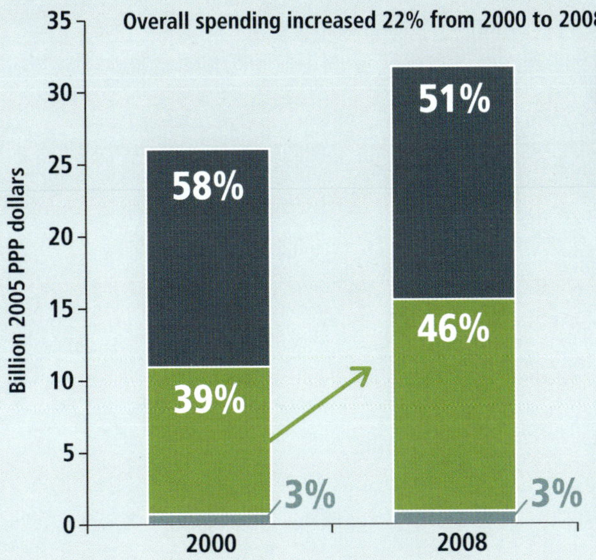

Main drivers of the $5.6 billion increase (2005 PPP $)	
$2.1 billion	China
$0.6 billion	India
$0.5 billion	United States
+/- $0.2 billion each	Brazil, Argentina, Iran, Japan, Nigeria, Russia

Source: Agricultural Science and Technology Indicators, *ASTI Global Assessment of Agricultural R&D Spending: Developing Countries Accelerate Investment* (Washington, DC: International Food Policy Research Institute, 2012).
Note: PPP = purchasing power parity.

Nienke Beintema is the head of the International Food Policy Research Institute's Agricultural Science & Technology Indicators (ASTI) initiative. **Gert-Jan Stads** is ASTI's program coordinator. **Keith Fuglie** is the chief of the Resource, Environmental, and Science Policy (RESP) Branch at the US Department of Agriculture's Economic Research Service (USDA-ERS) in Washington, DC. **Paul Heisey** is a senior economist in the RESP Branch of the USDA-ERS.

improving livelihoods—especially for poor farmers with very low productivity—means giving them better access to resources, technologies, and food. Regions that have lagged behind the agricultural technology frontier, such as much of Africa south of the Sahara, have remained mired in poverty and food insecurity. Countries in these regions could follow the examples of agricultural success stories like Brazil and China, which invested heavily in agricultural research, made critical reforms to policies and institutions, and tapped into international sources of agricultural technology to raise productivity, lower food prices, and stimulate economic growth. When a country's population shares broadly in these developments, it can have a major impact on lowering poverty and improving societal well-being. ■

Chapter 3

GREEN ECONOMY

Sustainable and Growing, but Also Food Secure?

Nitin Desai and Claudia Ringler

THE UNITED NATIONS CONFERENCE ON SUSTAINABLE DEVELOPMENT, held in Rio de Janeiro in June, was 2012's seminal event in the global dialogue on sustainable development. Called Rio+20, it was explicitly linked with the 1992 Earth Summit in Rio, the UN conference that put sustainable development on the global policy agenda.

Whereas the 1992 conference focused on sustainable development, the 2012 conference emphasized the related but distinct concept of a "green economy." Ultimately, Rio+20 was arguably short on specifics and commitments related to building a green economy, but there is potential for advancing a green economy agenda in several ways. This chapter explains the practical impact of the Rio+20 conference and the concepts of sustainable development and the green economy in the context of the past 20 years.

FROM AGENDA 21 TO THE GREEN ECONOMY

The 1992 Earth Summit sprang out of the work of the World Commission on Environment and Development, informally known as the Brundtland Commission, which was created in 1983. In 1987 the commission issued a report called *Our Common Future*, which famously defined sustainable development as "development that meets the needs of the present without compromising the ability of future generations to meet their own needs."[1] Later, the report elaborated: "the exploitation of resources, the direction of investments, the orientation of technological development, and institutional change are all in harmony and enhance both current and future potential to meet human needs and aspirations."[2]

The Earth Summit resulted in a voluntary, nonbinding action plan for sustainable development called Agenda 21. On the topic of sustainable agriculture and rural development, Agenda 21[3] proposed the following:

Nitin Desai is the former deputy secretary-general of the United Nations Conference on Environment and Development (1992) and former secretary-general of the World Summit on Sustainable Development (2002). **Claudia Ringler** is the deputy division director of the Environment and Production Technology Division at the International Food Policy Research Institute in Washington, DC.

Rio+20: Did It Move Us Forward?

MORGANE DANIELOU

For the agribusiness community organized under the Business Action for Sustainable Development, the 2012 United Nations Conference on Sustainable Development in Rio de Janeiro—known as Rio+20—had positive outcomes: recognition of the importance of agriculture for poverty reduction, acknowledgment of the imperative to sustainably intensify agricultural production, and awareness of the necessary mosaic of solutions. It affirmed that, within the UN system, the approach to the private sector is changing as it plays an ever more important role in delivering solutions.

Despite this, the outcome document adopted at the conference, called "The Future We Want," failed to instigate a true turning point toward greater global food security because it did not deliver a clear course of action. Rio+20 covered a wide range of issues so, by nature, cannot be considered an expert forum. For this reason, it did not—and perhaps could not be expected to—deliver the concrete policy guidelines needed to achieve food security. Instead, it mostly noted the efforts and developments that have taken place elsewhere.

In addition, Rio+20 lacked explicit entry points for leveraging the expertise of the private sector in delivering solutions. The initial attempt to focus on implementation gaps was valuable but did not find its way into the structure of the negotiations. It is a reflection of the nature of Rio+20 as a traditional intergovernmental negotiating process, which makes it valuable in its own right but limited in its ability to draw on external expertise or translate into multistakeholder commitments.

In contrast, the G8 and G20 Summits have been able to make influential decisions that with concerted effort have already led to action to increase food security. For example, the launch of the New Alliance for Food Security and Nutrition after the 2012 G8 Summit represents a clear commitment to change that benefits from the support of partnerships.

WHAT DID WE MISS?

In Rio, decisions on how to address implementation gaps could have been made and commitments reached on issues like increasing research capacity in developing countries or improving the delivery of extension services. However, because of its shape and nature, Rio+20 is not really the key forum on food security. It is fair to say that such decisions and commitments should be made elsewhere—in venues where the necessary experts are present and detailed principles and internationally accepted standards can be established, such as the Committee on World Food Security. These are the places where turning points can happen, as proven by the May 2012 adoption by this committee of the Voluntary Guidelines on the Responsible Governance of Tenure of Land, Fisheries and Forests in the Context of National Food Security.

The most concrete outcome of Rio+20 for food security will be the post-2015 agenda and the development of universal sustainable development goals, which aim to address the shortcomings of the UN Millennium Development Goals by focusing on eradicating poverty while protecting the environment. These goals will likely shape the way "sustainable" is defined in agriculture and draw on the expertise of those directly involved, including the private sector.

Morgane Danielou is the director of Communications at the International Fertilizer Industry Association in Paris, France.

- land planning, conservation, and rehabilitation,
- water for sustainable food production,
- conservation and sustainable agriculture,
- integrated pest management and control in agriculture,
- sustainable plant nutrition,
- a transition to more modern and efficient forms of energy in rural areas,
- diversification of farm and nonfarm employment,
- people's participation in decisions on sustainable development, and
- a review of agricultural subsidies and trade policies.

Arguably, the concept of sustainable development is particularly important in agriculture and food production, which are highly resource intensive and depend heavily on the integrity and viability of those natural resources. Agriculture and food production are also central to the social dimension of sustainable development, particularly as it relates to hunger and unmet food needs in developing countries.

Overall, the outcomes envisioned from Agenda 21 were an increase in food production and availability leading to a significant decline in hunger, a substantial reduction in poverty, improved conditions in critical ecosystems, and better integration of antipoverty efforts and natural resource management.

During the 20 years after the 1992 Earth Summit, acceptance of the sustainable development concept was consolidated in the global policy consensus. Nonetheless, there was arguably a lack of progress in implementing the programs agreed to at the summit. Trade, fiscal, and monetary policies often did not relate to or reflect sustainability concerns. Even if political will may have existed in ministries of environment and some other sectors, the public resources and the fiscal incentives and disincentives required were often not forthcoming from ministries of finance or higher levels of leadership.[4]

In 2008, in response to the perceived decline in public attention and gradual demotion of environmental issues in the policy priorities of governments, the United Nations Environment Programme (UNEP) advanced a new concept—the green economy—and pushed it hard for the next few years. UNEP defines a green economy as "one that results in improved human well-being and social equity while significantly reducing environmental risks and ecological scarcities. In its simplest expression, a green economy can be thought of as one which is low carbon, resource efficient, and socially inclusive."[5] At the same time other organizations, such as the World Bank and the Organisation for Economic Co-operation and Development, adopted a focus on the related concept of "green growth." Green growth implies not just that the world must bear the costs of environmental sustainability, but that it can benefit from it—through economic growth—while protecting the environment. As the World Bank described it, "inclusive green growth is the pathway to sustainable development."[6]

A green economy can be thought of as one that is low carbon, resource efficient, and socially inclusive.

WHY IS A NEW CONCEPT NEEDED?

After 1992, Agenda 21 and the suggested reorientation toward more sustainable development did not substantially improve poverty, hunger, or environmental outcomes. The absolute number of undernourished people, measured at about 1 billion during 1990–1992, has fallen over the past two decades, but it has stalled at around 870 million during the past five years—perhaps in large part because of the double burden of higher food prices and economic recession. Growing natural resource scarcity contributed to growing hunger and poverty as well.[7]

The Millennium Ecosystem Assessment reports that 15 of 24 critical ecosystem services are in decline (Table 1).[8] For example, although deforestation rates have declined in the Caribbean and East, South, and West Asia, they have increased over the past 20 years in other regions, particularly parts of Latin America, much of Africa, and Southeast Asia.[9] The global freshwater fishery sector is in rapid decline in terms of both the biodiversity and quantity of aquatic species. The ocean commons remain highly exploited, even overexploited, and the outlook appears grim.[10]

Unsurprisingly, land scarcity is rapidly worsening,[11] and degradation continues apace. Whereas past efforts to combat land degradation have focused on drylands, the highest rate of land degradation over the past three decades has occurred in the humid and subhumid tropics.[12] The "land grab" phenomenon can be seen as a signal of weak land governance and management. Research on land

TABLE 1 Global status of ecosystem services evaluated in the Millennium Ecosystem Assessment

Ecosystem service	Status
Food	
Crops	Enhanced
Livestock	Enhanced
Capture fisheries	Degraded
Aquaculture	Enhanced
Wild foods	Degraded
Fiber	
Timber	Both enhanced and degraded
Cotton, hemp, silk	Both enhanced and degraded
Wood fuel	Degraded
Genetic resources	Degraded
Biochemicals, natural medicines, pharmaceuticals	Degraded
Fresh water	Degraded
Air quality regulation	Degraded
Climate regulation	
Global	Enhanced
Regional and local	Degraded
Water regulation	Both enhanced and degraded
Erosion regulation	Degraded
Water purification and waste treatment	Degraded
Disease regulation	Both enhanced and degraded
Pest regulation	Degraded
Pollination	Degraded
Natural hazard regulation	Degraded
Spiritual and religious values	Degraded
Aesthetic values	Degraded
Recreation and tourism	Both enhanced and degraded

Source: Millennium Ecosystem Assessment, *Ecosystems and Human Well-Being: Synthesis* (Washington, DC: Island Press, 2005), Table 1.
Note: Status indicates whether the condition of the service globally has been enhanced (if the productive capacity of the service has been increased, for example), degraded, or both.

use change and land use planning has advanced, but some argue that there has been little implementation on the ground. Shifting land toward more sustainable uses can be a slow process fraught with institutional and governance difficulties, including the continued conversion of natural land resources for human use. Lack of secure land use rights remains a challenge to sustainable rural development and human well-being in parts of the developing world.[13]

Water scarcity and degradation are also increasing and have been attributed to economic and population growth, poor water management, and the impacts of climate change.[14] Water use efficiency has improved—in part thanks to continued advances in seed technologies and management practices—but much of the improvement was designed to counteract economically damaging water shortages rather than to proactively support sustainability goals.[15]

Little improvement has been made on other policy elements of Agenda 21. Subsidies for agricultural inputs and output may be even more prevalent. The developed countries are maintaining subsidies that encourage overproduction and distort trade.[16] Many developing countries have initiated their own large subsidy programs for water, energy, and fertilizers, even as these become increasingly fiscally unsustainable because of higher prices and greater need.[17] On trade distortions, some countries have scaled back tariffs, but the failed Doha Round of trade negotiations is a stark reminder that all is not well on the agricultural trade front. The failure may have negative consequences for global environmental sustainability, as continued trade distortions prevent food crops from being grown where comparative biophysical advantages, such as rainfall or fertile land, are greatest.[18]

At the same time, it appears that CGIAR, a consortium of international agricultural research centers, has underinvested in ecosystem-compatible agricultural systems. Some argue that such investment has remained limited because it has proven difficult to assess the full costs and benefits of environmental services and sustainability and because CGIAR sees a focus on seeds and technologies as

Green and Greener: Toward Sustainable Agriculture

SYLVIE LEMMET

WHAT HAPPENED IN 2012?

Food security is a pressing global concern. At the 2012 United Nations Conference on Sustainable Development (Rio+20), participants reaffirmed their commitment to enhancing food and nutrition security for present and future generations by developing strategies at all levels that align with the Five Rome Principles for Sustainable Global Food Security.[1] To revive agriculture, it was agreed that investments are necessary in these areas: sustainable agricultural practices, rural infrastructure, storage capacities and related technologies, research and development on sustainable agricultural technologies, strong agricultural cooperatives and value chains, and urban-rural linkages.

A green economy is one that improves human well-being and social equity while significantly reducing environmental risks and ecological scarcities; the participants of Rio+20 recognized that such an economy is necessary to achieve sustainable development and poverty reduction. In its *Towards a Green Economy* report, the UN Environment Programme (UNEP) argued that "greening" agriculture will require investment in soil-fertility management, sustainable water use, crop and livestock diversification, biological plant and animal health management, mechanization levels, storage facilities (especially for small farms), and supply chains—both upstream and downstream—for businesses and trade.[2] The aggregate global cost of these and other investments and policy interventions required to transition to green agriculture is estimated at US$198 billion per year from 2011 to 2050.

As a follow-up to Rio+20, UNEP is currently working with countries worldwide, responding to demands for technical assistance and capacity building to enable a green economy in various sectors, including food and agriculture. Through its partnership with other agencies, UNEP aims to strengthen the capacity of governments and other stakeholders to manage the transition to socially inclusive, resource-efficient, low-carbon economies and to provide a springboard for action on commitments made at the conference.

WHAT ARE THE NEXT STEPS?

With a greener agriculture sector, UNEP stated that a growing and more demanding world population could be nourished with the estimated 3,200 kilocalories per person per day made available up through 2050. But additional research to disaggregate this scenario at regional and national levels is required. Similarly, given the rapid depletion of ecosystems and natural habitats resulting from farming, there is a need to assess the value of ecosystem services and their role in food production, conservation of vital processes such as maintaining clean water, and waste decomposition.

There is also a need to economically quantify the value of diminishing natural resources and environmental commodities as well as to explore the relationship between changes in ecological factors and food price volatility. Finally, while current research covers the economic value of food stocks that are either lost or wasted, there is a lack of information on their true value, which incorporates the value of the natural resources, such as water, fuel, and fertile soil, embedded in producing these food stocks.

Sylvie Lemmet is the director of the United Nations Environment Programme's Division of Technology, Industry, and Economics, which is based in Paris.

the best way to support rural livelihoods, incomes, and overall environmental outcomes.[19] Moreover, national governments have invested little in ecosystem service improvements, with some exceptions, such as China, which may be responding to the obvious adverse consequences of environmental degradation.[20]

RIO+20 AND THE POTENTIAL FOR ACTION

At the Rio+20 conference, the green economy remained a controversial subject. Some developing countries and civil-society groups were concerned that a new valuation of natural resources would serve as an excuse for grabbing the resources of poor countries and people. There was fear that the terminology would be hijacked by the private sector to advance high-input intensification and industrial agriculture.[21]

The acceptance of green economy ideas in the Rio+20 outcome—the UN General Assembly resolution known as "The Future We Want"—was thus lukewarm. The 19 paragraphs devoted to the green economy seem to emphasize qualifications to the acceptance as much as the acceptance itself: "we consider green economy in the context of sustainable development and poverty eradication as one of the important tools available for achieving sustainable development and that it could provide options for policymaking but should not be a rigid set of rules."[22] There is no clear commitment on the greening of economic policies such as agricultural or energy subsidies.

Ties That Bind Energy, Food, and Agriculture

EUGENIO DÍAZ-BONILLA

Agriculture and energy have always been interrelated, but the current civilization's dependence on fossil fuels has redefined the relationship between them to one focused on agriculture and energy costs, in terms of producing, processing, transporting, and storing agricultural and food products. The Green Revolution occured at a time when energy costs were relatively lower and evolved in a context where greenhouse gas emissions were not considered a constraint. At a macroeconomic level, sharp increases in oil prices have affected disposable incomes and generated recessions, creating impacts on agriculture.[1] Recently, the links between energy and agriculture have expanded and become more complex.

First, biofuel mandates increased the demand for agricultural products as fuel inputs, and developments in the nonfood energy sector now clearly affect food markets because of the difference in the size of the food energy market versus the nonfood energy market. If all the food energy needed for human beings to function and all the nonfood energy used by the world to operate is calculated in a common measure (joules, for example), the latter amount is about 16 to 18 times higher than the former. Second, the correlation of world prices of oil and agricultural commodities has become stronger in recent years because of financial investments in commodities. Finally, another link between agriculture and energy comes from climate change impacts attributed to fossil fuel use and emissions.

These links need to be considered in the current context of very high real oil prices, which began climbing in the first half of the 2000s. Although prices declined from their monthly peak in March 2012, they were very high throughout the year and are projected in the next decades to be even higher than previous peaks, according to estimates by the International Energy Agency. The first period of high prices from the mid-1970s to the mid-1980s was sustained by strong world growth and geopolitical events, but it ended in the mid-1980s because growth softened, debt crises occurred in developing countries, and energy saw numerous technological innovations. The current high prices make the future similarly uncertain. While global economic prospects are again deteriorating, new technologies like shale gas and tight oil are reducing energy prices in some large markets with important potential implications for agriculture, from fertilizer production to global emissions.

So, will we face a scenario similar to the one seen in the mid-1980s, when technological developments in energy and depressed global macroeconomic conditions led to a collapse in energy prices? Or, is the world moving toward a situation of sustained real energy prices at levels not yet experienced in history? To answer these questions in a way that incorporates and potentially benefits agricultural production, food security, and poverty reduction, a more systematic and integral view of the complex links between energy and agriculture is required.

Eugenio Díaz-Bonilla is a visiting senior research fellow in the Markets, Trade, and Institutions Division of the International Food Policy Research Institute, Washington, DC.

Discussions on agriculture and food at Rio+20 added little to what had been agreed upon at the Earth Summit 20 years earlier, as is clear from the following statement from "The Future We Want":

> We reaffirm the necessity to promote, enhance and support more sustainable agriculture, including crops, livestock, forestry, fisheries and aquaculture, that improves food security, eradicates hunger and is economically viable, while conserving land, water, plant and animal genetic resources, biodiversity and ecosystems and enhancing resilience to climate change and natural disasters. We also recognize the need to maintain natural ecological processes that support food production systems.[23]

Nonetheless, some elements in the outcome of the Rio+20 conference could change how sustainable development is handled at the international level. First, ecological concerns could be mainstreamed into the central elements of economic policy. "The Future We Want" is less precise than its proponents wanted. But there is a window of opportunity to develop green economy ideas further in the invitation to the United Nations, donors, and international organizations to provide toolboxes, best practices, models, good examples, and methodologies for policy evaluation.[24]

Second, the conference resulted in a commitment on a process to formulate Sustainable Development Goals (SDGs). The Millennium Development Goals (MDGs), which emerged from the UN's Millennium Summit, focused attention largely on quantifiable targets that touched individual lives, such as halving poverty, reducing maternal and child mortality, and enhancing girls' education. Although one goal spoke about sustainable development in general terms, the more collective dimension of resource management got lost. Now the UN system will move ahead to complement—but not replace—the MDGs with goals and targets that include a stronger sustainability agenda that would build directly on Agenda 21. Areas suggested for SDGs include combating poverty, changing consumption patterns, promoting sustainable human settlement development, preserving biodiversity in forests, protecting oceans and water resources, advancing food security, and promoting sustainable energy, including from renewable resources.[25] The very fact that the mandate for the SDGs has come from Rio+20 might strengthen the hands of those who want the resource dimension to play as large a role in defining these global goals as poverty reduction, health, and education.

Third, a new high-level forum may offer leadership on green growth. Although the institutional outcome is still uncertain and must be negotiated further in the General Assembly, "The Future We Want" specifies that it would be a body with universal membership (that is, it would include all UN member states) aimed at, among other things, providing political leadership, advancing the integration of the three dimensions of sustainable development, and enhancing systemwide coherence and coordination.[26]

The commitment to agree on Sustainable Development Goals may provide an impetus for action.

Also showcased in Rio was the UN secretary-general's Sustainable Energy for All initiative, a multistakeholder partnership. It seeks universal access to modern energy services, a doubling of the rate of improvement in energy efficiency, and a doubling of the share of renewable energy in the global energy mix by 2030. In some ways it is a response to the heightened concern about climate change. Each Intergovernmental Panel on Climate Change report since 1990 has been more definite about the anthropomorphic causes of climate change and more alarming about impact than the previous one.[27] But energy is so central to all sectors that climate action must encroach on core areas of economic policy in order to change how energy is being produced and consumed, and thus how economies are run, to advance sustainable development and green growth.

At Rio+20 and beyond, policy discussions about the green economy and sustainable development have occurred in parallel with developments in the

Advances in efficiency must go hand in hand with improvements in the lives of poor and food-insecure people.

private sector. Some private companies have started to promote sustainability and stewardship programs not only to support their own longer-term success, but also to attract more shareholders.[28]

Last but not least, environmental sustainability has made its way into school curricula and public thinking, chiefly in developed countries—as evidenced by a recent opinion survey carried out in Germany on the topic of global food security[29]—but also incipiently in developing countries. Moreover, consumer and environmental civil society groups have increased pressure for change toward sustainability. The results of this increased social engagement could be seen in the scale and impact of civil-society activity at Rio+20.

WHERE TO GO FROM HERE

The integration of sustainability with development, green policies with growth, and resource management with poverty eradication is not ensured. The commitment to agree on Sustainable Development Goals may provide an impetus for action. The SDGs provide an opportunity to integrate resource management and poverty eradication, a task that seems to have been largely ignored as governments and aid agencies have moved resources to antipoverty programs targeted directly to certain groups. It remains to be seen, however, if and how the SDGs will eventually make a difference for the environment and the poor. The potential for trade-offs between these goals was well expressed by Agenda 21:

> While managing resources sustainably, an environmental policy that focuses mainly on the conservation and protection of resources must take due account of those who depend on the resources for their livelihoods. Otherwise it could have an adverse impact both on poverty and on chances for long-term success in resource and environmental conservation. Equally, a development policy that focuses mainly on increasing the production of goods without addressing the sustainability of the resources on which production is based will sooner or later run into declining productivity, which could also have an adverse impact on poverty.[30]

Even if governments are now asked to reorient policies and investments voluntarily toward green growth objectives (and future SDG targets), the actual achievement of these objectives will likely continue to be subject to the vagaries of economic growth and environmental stress.

Another factor that could energize implementation of green growth is a closer integration with the agenda emerging from the climate change negotiations. Agriculture, which is predicted to be seriously affected by the anticipated temperature and precipitation changes, is an important part of this agenda. If governments and international organizations step up their efforts at adaptation activities, they could create a window of opportunity for pursuing the green growth agenda. Along those lines, the secretary-general's Sustainable Energy for All initiative can help focus attention on the land-water-energy nexus that is central to a green economy.

There are other global programs with which a program for sustainable agriculture could find synergies. One example is biodiversity protection. Agriculture depends on biodiversity, not just for the wild ancestors and relatives of cultivars, but also for soil biota, natural predators of pests, and pollinators. Similar synergies could arise if the growing global concern about freshwater scarcity leads to stronger commitments by governments and greater transnational cooperation.

A more difficult area for policy research and advocacy is related to the invitation to assist UN member states with toolboxes, best practices, and methodologies for policy evaluation that would allow green economy concerns to shape public policy. In the context of agriculture, such an effort would involve re-examining the perverse subsidies on inputs and outputs—issues that have so far found no policy consensus among national governments.

A word of caution is necessary. Worsening natural resource scarcity could provide an added push to move toward more efficient resource use—but any advances in efficiency must go hand in hand with improvements in the lives of poor and food-insecure people. Additionally, economic growth, financial stability, and fiscal health will likely remain fragile for many more years given the present state of the world economy.[31] And weak political will and limited fiscal capacity may stand in the way of effective action by governments. Responsibility for achieving a green economy and eradicating poverty may thus lie partly in the hands of civil society and the private sector, which will need to be provided with the incentives and information they need to ensure that their actions support poverty reduction, food security, and green growth. ■

Chapter 4

WOMEN IN AGRICULTURE

Closing the Gender Gap

Ruth Meinzen-Dick and Agnes Quisumbing

THE ROLE OF GENDER IN AGRICULTURE GAINED GROWING ATTENTION among researchers, aid donors, and policymakers in 2012. As new knowledge on the role of gender in agricultural productivity emerges, agricultural programs can use this knowledge to improve outcomes. But more knowledge and better outcomes will depend on a stronger evidence base on gender in agriculture, programs that do more to address the particular needs of women as well as men, and more vigorous engagement with women's groups as full partners in agricultural development.

SO FAR: GROWING ATTENTION TO GENDER IN AGRICULTURE

Two recent flagship reports—the Food and Agriculture Organization's (FAO's) *State of Food and Agriculture 2010–2011* and the World Bank's *World Development Report 2012*—turn their attention to gender issues in agriculture. Emphasizing women's contributions to agriculture in developing countries, *The State of Food and Agriculture 2010–2011* highlights the need to close the gender gap in access to agricultural resources, education, extension, financial services, and labor markets; to invest in labor-saving and productivity-enhancing technologies and infrastructure to free women's time for more productive activities; and to facilitate women's participation in flexible, efficient, and fair rural labor markets.[1]

The *World Development Report 2012* stresses that gender equality can lead to productivity gains, that women's increased control of household resources can improve outcomes for the next generation, and that empowering women as economic, social, and political actors can result in more representative decisionmaking.[2] The report also identifies areas where policy can help close the gender gap: addressing excess deaths of girls and women; improving girls' education; equalizing access to economic opportunities and reducing productivity

Ruth Meinzen-Dick is a senior research fellow in the Environment and Production Technology Division of the International Food Policy Research Institute (IFPRI) in Washington, DC. **Agnes Quisumbing** is a senior research fellow in IFPRI's Poverty, Health, and Nutrition Division.

gaps between women and men; giving women a stronger voice in households and societies; and limiting the transmission of gender inequality across generations.

Attention to gender in agriculture is not new, but in the past it has often been limited to a few specialized programs targeting women or "mainstreaming" efforts that embed attention to gender within programs, with too little follow-through. That seems to be changing. A number of key development agencies and donors are drawing on gender analysis in their programming, targeting by gender, and building in accountability. For example, FAO's 2012 gender strategy commits to allocating 30 percent of operational budgets to interventions targeted to women and to disaggregating data in all FAO statistical databases by sex. The Bill & Melinda Gates Foundation's policy on gender-responsive agricultural programming is summarized as "Know Her, Design for Her, Be Accountable to Her."

Development programming is now moving from gender-blind programs that ignore gender differences, to gender-aware programs that recognize the different needs of men and women.

Similarly, the new gender strategy of the US Agency for International Development was accompanied by the creation of the Women's Empowerment in Agriculture Index, which will be used to measure progress toward inclusive agricultural growth in all 19 countries where the agency's Feed the Future program is in place. Building on the momentum in the development community, and drawing upon two decades of gender research in CGIAR, the Consortium Board issued a gender strategy in November 2011, which integrates gender analysis in all Consortium research programs. All of these initiatives not only address gender equality and the specialized needs of women in agriculture, but also create accountability for delivering on these commitments.

Development programming is now moving from gender-blind programs that ignore gender differences, to gender-aware programs that recognize the different needs of men and women, and even to gender-transformative projects that seek to promote more gender-equitable relationships. As part of the Gender, Agriculture, and Assets Project supported by the Bill & Melinda Gates Foundation, IFPRI researchers are studying how projects that pay more attention to gender differences in rights, resources, and responsibilities are better able to achieve their development objectives. For example, in the first phase of a recent dairy development project in Mozambique, dairy cows were distributed to households, but training and control of dairy income were directed to men. Preliminary study results suggest that because dairy farming requires substantial inputs of women's labor, the program's focus on men created tension within households. Project designers and implementers took notice when a woman deliberately starved a cow distributed by the project because she believed the project rewarded men but failed to recognize her input. The project was eventually modified to allow two household members—typically a man and a woman—to attend training on dairy management. The recently funded second phase of the project deliberately paid greater attention to gender issues.

Other projects have taken a more gender-transformative approach. In two states of India, where women traditionally do not inherit land, Landesa (known in India as the Rural Development Institute) is allocating small plots of land to landless households and attempting to change the rules of the game for future generations. First, plots allocated to dual-headed households are titled jointly in the names of the husband and wife, and female-headed households and daughter-only households are given priority for land allocation. Moreover, all sons and daughters are listed as co-inheritors in the land title—an innovation for rural India, where girls are disadvantaged in terms of land ownership. Working with the Ministry of Women and Child

Getting Gender Roles Right: A Success Story in Mozambique

JEMIMAH NJUKI AND ELIZABETH WAITHANJI

Although many smallholder farmers are women, women are rarely consulted during the design phase of farm-focused development interventions. When programs do not take into consideration the needs and concerns of 100 percent of their beneficiaries, they often run into problems. For example, in 2006, after years of civil conflict had taken its toll in Mozambique, the Smallholder Dairy Development Program funded by the US Department of Agriculture and implemented by Land O'Lakes sought to rebuild the country's dairy industry to meet market demand, raise incomes, and increase crop yields by using working animals. The intervention saw mixed results in its early years, and the positive outcomes were not always benefiting all members of a household. So, in an effort to engage both men and women, the program started requiring two household members to be trained in dairy production instead of just one.

This did not always lead to the inclusion of women, however, because often a man and his son would be trained in exchange for the cows. Soon farmers began returning some cows and others died. Why? Because the women in those households refused to feed them. Since they did not co-own the cows or have any control over the resulting income, the women farmers saw no reason to add to their own workload.

Enter the Gender, Agriculture, and Assets Project (GAAP). Led by IFPRI and the International Livestock Research Institute, GAAP works with agricultural development practitioners to analyze gender roles and account for gender differences within interventions and impacts. The GAAP team collected data and facilitated focus-group discussions on gender roles in dairy production, cattle ownership, labor, marketing, and income management. Together with a Mozambique program team, they developed interventions to address the gender gaps and designed a monitoring and evaluation system. Ultimately, a new game plan emerged: (1) register the cows given to each household in the name of at least one male and one female adult, (2) organize and train groups of farmers on gender relations and promote women's leadership, (3) register both male and female adults in the marketing cooperative, and (4) include women in the cooperative management committee. The results? More women own livestock. More women are involved in making decisions about managing that livestock. And more women market milk and make use of income from its sale.

These valuable lessons about gender integration have led to wider discussions between Land O'Lakes International Development's managers about the role of gender in their programming, and the results have significantly influenced the second phase of funding for the program, which began in 2012.

Jemimah Njuki is the global coordinator for the Pathways program on women in agriculture at the Cooperative for Assistance and Relief Everywhere (CARE), based in Dar es Salaam, Tanzania. **Elizabeth Waithanji** is a postdoctoral scientist at the International Livestock Research Institute in Nairobi, Kenya.

Development, the project is creating girls' and boys' groups that incorporate a land rights curriculum and is holding community conversations, including both men and women, on girls' inheritance rights.

Initiated in 2002 and now in its third phase, the Targeting the Ultra Poor program of BRAC transfers livestock to women in ultra-poor households, providing livelihood assistance and training in income-generating activities. Since 2007, CARE–Bangladesh's Strengthening the Dairy Value Chain project has included activities to empower women and challenge barriers to women's participation in the dairy value chain. The project has helped women overcome barriers to mobility by moving milk collection points to villages, increasing the number of local input supply points, and training a cadre of female livestock health workers. Preliminary study results suggest that through group savings schemes, many women have been able to save enough money to buy a dairy cow. IFPRI researchers are working with project designers and implementers to assess the impact of these projects

Indexing Women's Empowerment

EMILY HOGUE AND CAREN GROWN

"Something that can be measured can be changed."

—Hillary Rodham Clinton, Former US Secretary of State

A great deal of international development goals can be tangibly measured—in crop yields or new jobs or reductions in child mortality, for example—while others prove much harder to quantify. Women's empowerment, for example. How do you define it, measure it, collect those measurements, and analyze them? The US Agency for International Development (USAID) recently had the opportunity to answer these questions as it developed the Women's Empowerment in Agriculture Index (WEAI) with partners from IFPRI and the Oxford Poverty and Human Development Initiative. The tool was developed for the US government's Feed the Future initiative to reduce poverty and food insecurity.

The index is a significant innovation in its field that measures multiple indicators of empowerment, and generates "scores" that can be compared over time. It is the first tool of its kind. Piloted in 2011 and launched in February 2012, the US government is using the index to track change in women's empowerment that occurs as a direct or indirect result of Feed the Future interventions in targeted geographic zones within the initiative's 19 focus countries.

Data for the WEAI will be collected every two years in all 19 countries, and baselines were collected in 2011 and 2012. USAID and partners will conduct data analyses to understand the relationships among empowerment, livelihoods, and food security, as well as relationships among the various components of the index. Feed the Future will also use the WEAI for impact evaluations of distinct projects to examine the effectiveness of different approaches and how they impact women and men. Through IFPRI, Feed the Future has selected and began funding four dissertation grants for research that will build evidence on women's empowerment through diverse methodologies and substantive areas. All of this analysis will help project leaders refine the WEAI for improved practicality and broad utility.

Many development partners have expressed interest in using the index for their programs, and USAID and IFPRI are working to develop tools and guidance to replicate it beyond the 19 focus countries. What started as a fairly modest effort to develop a monitoring tool for Feed the Future has greatly exceeded expectations and provided the development community with a robust and accessible instrument to tackle one of the most complicated development challenges. While just a first step toward improving learning and programming in this critical area, the creation of the index signifies the commitment of the US government to prioritizing women's empowerment as an essential development outcome it will measure and strive to achieve.

Emily Hogue is the team leader for Monitoring and Evaluation in the Bureau for Food Security at the US Agency for International Development (USAID) in Washington, DC. **Caren Grown** is a senior gender adviser at USAID.

on the gender asset gap and broader development objectives.

Gender-sensitive development projects have also learned to use local resources and institutions. A local adaptation of the homestead food production program implemented by Helen Keller International in Burkina Faso from 2009 to 2012 involved training grandmothers as sources of nutrition advice, given the important role of older women in the local culture. In Uganda, the initial effort to propagate orange sweet potato vines developed by HarvestPlus from 2007 to 2009 relied on social networks and farmers' groups, and extension messages on the health benefits of consuming orange-fleshed sweet potatoes were deliberately targeted to women.

Although donor priorities are a clear signal of attention to gender issues, this is far from being a donor-driven process: projects that pay attention to gender differences on the ground are also projects that are more likely to succeed.

WHAT NOW? MORE IS NEEDED

Political will and project interventions to address the needs of women in agriculture and contribute to gender equality are increasing. However, additional efforts are needed to lock in these gains so that attention to gender becomes an integral part of agricultural and rural development—not merely a development trend that passes. More needs to be done to build the evidence base on gender in agriculture, strengthen women's assets, and engage with women's groups as real partners in development.

Build the Evidence Base

The evidence marshaled by *The State of Food and Agriculture 2010–2011* and *World Development Report 2012* has helped build the political will to address gender in agriculture. But examination of the statistical appendixes in *The State of Food and Agriculture* shows that this evidence is still fragmentary, especially in agriculture. National-level agricultural statistics too often fail to report whether farmers and agricultural laborers are men or women. Even most agricultural surveys treat households as single units rather than disaggregating data by sex. What passes for gender analysis is limited to comparisons of "male-headed" and "female-headed" households, ignoring the majority of women who live in households with males and likely understating gender differences in productivity.[3] Figures on the extent of women's ownership of and access to land and other assets often conflict because of different ways of collecting information. Moreover, the conventional wisdom on gender is often based on data that are drawn from extremely small samples or are already outdated.

Fortunately, there are now methods that can be applied to measure the gender gap in assets. One such method was used in a recent study based on statistically representative datasets in Ecuador, Ghana, and Karnataka State, India.[4] The Wikigender initiative of the Organisation for Economic Co-operation and Development's Development Centre (www.wikigender.org) and the World Bank's Gender Data Portal (http://datatopics.worldbank.org/gender) represent important steps toward building a systematic evidence base and provide resources for policymakers and program designers on a range of gender issues. This kind of evidence is needed to identify the scope of the challenges that women face in agriculture and how these challenges vary across regions, time, and socioeconomic status.

More needs to be done to build the evidence base on gender in agriculture, strengthen women's assets, and engage with women's groups as real partners in development.

Women's empowerment and gender inequality are typically measured at an aggregate country level, using available data such as education rates or percentage of women in parliament. These indicators are important, but they are not directly related to agriculture. In contrast, the Women's Empowerment in Agriculture Index collects new information by interviewing men and women themselves about their participation in decision-making about agriculture, control over productive resources, control of income, leadership, and time allocation. This information is used to construct indicators of five domains of empowerment for women, as well as gender parity between men and women in the same household (Figure 1).[5] While this index can be used to track progress over time, it also has a more immediate use as a diagnostic tool. It can identify the key areas in which women (and men) lack empowerment so that programs can focus on those areas. It also draws attention to how agricultural programs can strengthen women's empowerment, such as by contributing to women's leadership through group-based approaches, or weaken it, such as by creating an excessive workload. One could argue, for example, that the dairy program in Mozambique described earlier would not have involved only men if program designers had thought through how it would affect the five domains of empowerment.

Better evidence on what works to empower women is needed to help guide development practice, to screen out ineffective programs, and to strengthen the case for channeling resources to effective ones. Rigorous quantitative studies are important, but they should be complemented by qualitative studies that can both identify key questions for quantification and help explore the dynamics of how programs work (as is being done in the Gender, Agriculture, and Assets Project cited earlier). Including women and men directly in monitoring and evaluation enhances both understanding and accountability, as implied in the "be accountable to her" pillar of the Bill & Melinda Gates Foundation's gender in agriculture strategy.

Moreover, gender relations do not change overnight. Long-term studies are needed to assess whether gender-transformative approaches are having effects that may not be apparent in the short term. An IFPRI study of the impact of vegetable and fish pond technologies in Bangladesh illustrates the value of longer-term impact assessments. Although early adopters of fish pond technologies had greater gains in income and food consumption than later adopters, the vegetable technologies were disseminated throughout the villages by women's groups and led to greater positive effects on the nutritional status of women and children, as well as greater gains in women's assets—impacts that would not show up in the short term.[6]

Strengthen Women's Assets

Redressing the gender gap in assets is key to improving agricultural productivity. Policies and programs that strengthen women's assets are likely to have long-term effects because they not only increase women's ability to adopt improved technologies or engage in more remunerative livelihood strategies, but also contribute to women's empowerment in the household and the community. Key assets in this regard include tangible ones such as land, water, tools, and technologies, as well as less tangible—but no less important—ones such as financial, human, and social capital.

LAND. Joint titling and reform of inheritance laws are two major policy reforms being employed to strengthen women's property rights. For example, land is being jointly certified in the names of both husbands and wives in Ethiopia, and reforms to the Hindu Succession Act in India are allowing daughters to inherit land. However, legislative reforms alone are not enough to affect the lives of rural women because customary land rights and inheritance practices often remain strong. Full implementation of any legislative reform is crucial. Legal literacy programs are needed to make both communities and the implementing agencies aware of the reforms and the rights that women can claim. In Ethiopia, for example, putting women on the local land administration committees has helped to increase women's awareness of their rights and the likelihood that they will register their land.[7] Enlisting male elders to support women's land rights can play a critical role in establishing social legitimacy for reforms.

WATER. Reliable access to clean water is especially important to women, who are often responsible for obtaining water for domestic use and for plants and animals under their care. Although domestic water supplies and irrigation systems are often developed separately, some studies suggest that creating multiuse water systems helps women meet their domestic as well as productive water needs.[8] Homestead-scale systems such as wells and water harvesting require substantial private investment; community-scale systems can transport water from longer distances and may offer technical efficiency gains. However, community-level systems require effective organizations. Women's participation in decisionmaking in community water systems is crucial to ensure that their particular needs and priorities are addressed.

TOOLS AND TECHNOLOGIES. A gender-aware approach to closing the gender asset gap in agriculture requires ensuring that women have access to and control over the tools and technologies that men already have. It is thus important to know whether women are less likely to have the tools because of cost or cultural factors: is a woman unable to afford farm equipment or communications technology, or is it socially unacceptable for her to have or use it? A gender-transformative approach requires asking whether technologies are designed to meet women's needs and whether women are involved in the innovation systems, both as clients and as providers of innovations. For example, women and men often have different trait preferences for crop or livestock varieties; participatory plant or animal breeding programs that involve women can take these differences into account.[9] Ergonomically designed equipment for women can reduce strain and make their labor more productive. Many agricultural research and development programs focus on the period from planting to harvest and neglect postharvest processing, which is more likely to be women's domain. Improvements in processing can reduce food losses, increase incomes, improve nutritional

FIGURE 1

WOMEN'S EMPOWERMENT IN AGRICULTURE

The Women's Empowerment in Agriculture Index measures women's empowerment by assessing five domains. It also compares women's scores in these five areas with those of other men in their household to measure gender parity

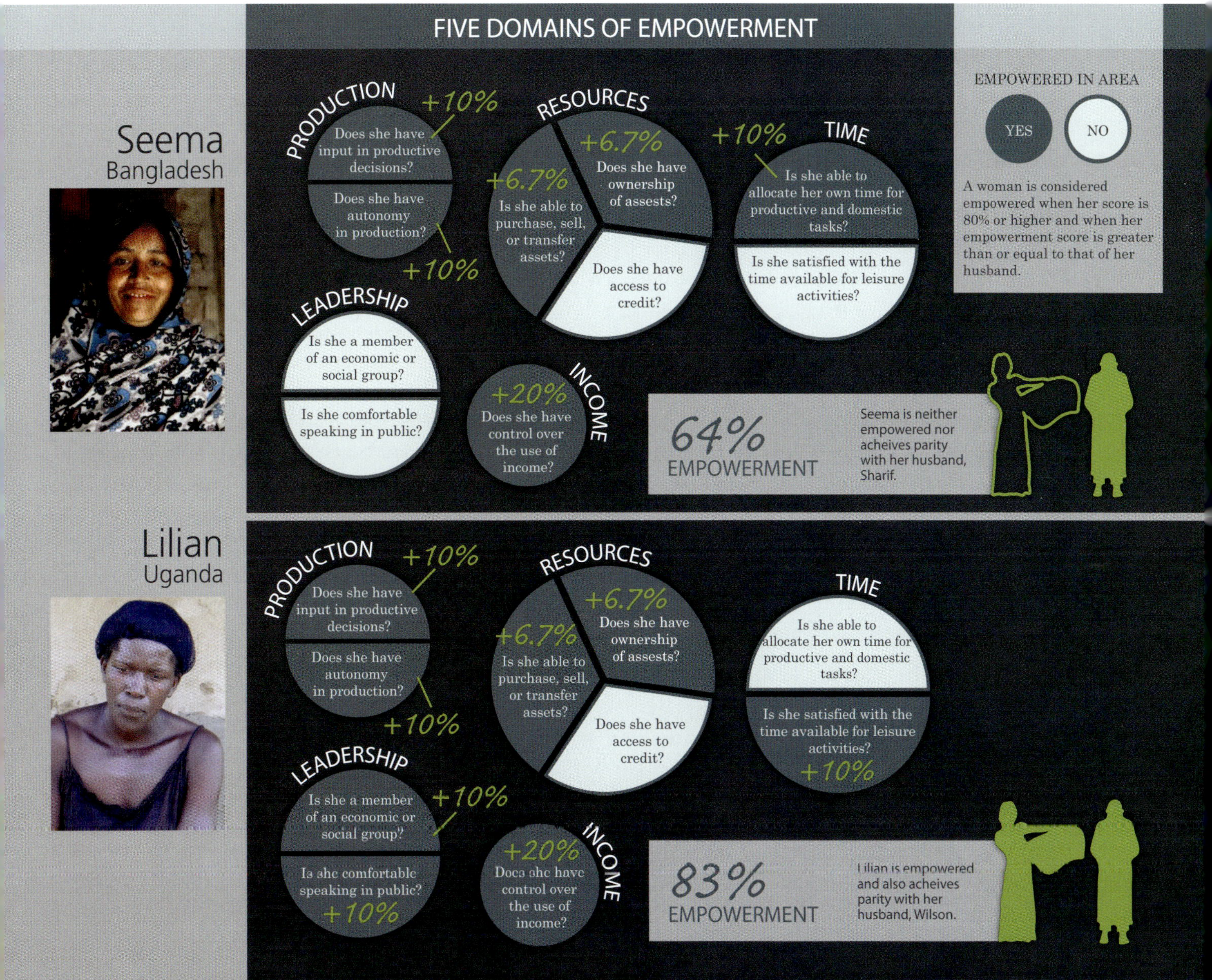

Source: United States Agency for International Development, International Food Policy Research Institute, and Oxford Poverty and Human Development Initiative, Women's Empowerment in Agriculture Index, http://www.ifpri.org/publication/womens-empowerment-agriculture-index, accessed November 25, 2012.

Attention to gender in agriculture is not new, but it has not always been acted upon.

content, and save labor—all of which are of special concern to women.

FINANCIAL CAPITAL. Microfinance programs have often successfully made savings and credit available to poor women, but women should not be locked into microfinance alone; a ladder of finance is also required. In addition, women need more than credit: ways of making and receiving payments, such as through mobile phones, allow women to participate in markets for inputs and produce. These mechanisms can be particularly important in societies that restrict women's mobility, because women there must rely on male family members or friends to take their produce to market.

HUMAN CAPITAL. Much attention is rightly given to increasing girls' enrollment and retention in school. Later in life, however, women also need greater access to agricultural extension and advisory services; women have consistently lower access to extension than men.[10] More gender-equitable extension systems would not only recognize women as farmers, but also address their needs for information, technology, and market access. They should employ female extension agents or alternative delivery mechanisms such as group-based approaches and experiential techniques that are specifically designed to be gender equitable, depending on what is most effective.

SOCIAL CAPITAL. Both governments and nongovernmental development organizations have embraced group-based approaches as a way to reduce the cost of delivering services to many individuals and make program outreach more cost-effective. Participation in groups can also be empowering. For women, the opportunity to get out of the house and meet with other women, create solidarity, and build confidence to speak in public can increase their bargaining power within their households as well as with outsiders. Informal social networks and kinship ties can also be important sources of information, influence, and assistance. But, as with other assets, men often have stronger social capital than women have. Ensuring that the formal rules and informal practices of groups do not discriminate against women can help redress this gap. For example, membership rules should allow multiple household members to participate instead of just the head of household; meetings should be scheduled where and when women can attend and should be conducted so that women feel they can be heard.

Engage in Real Partnerships

Ultimately, translating political will into gender-equitable agricultural development requires forging partnerships with women's organizations so that they have voice and agency. Instead of having to anticipate women's needs, programs should make it possible for women to identify their own needs, place demands on service providers, and increase accountability of programs. This approach should be adopted not only by government and nongovernmental organizations, but also by private input and information suppliers, land administration services, and marketing agents.

This is not to suggest ignoring men: indeed, enlisting men's support and ensuring that their needs and concerns are also addressed are crucial for the advancement of both men and women. Effective partnerships have the potential to transform how women are perceived. Instead of being seen as victims of inequitable conditions or beneficiaries of programs, women can be seen as key actors in implementing programs and achieving sustainable development, food security, and poverty reduction.

CONCLUSIONS

Attention to gender in agriculture is not new, but it has not always been acted upon. The accumulating evidence shows that empowering women is not only important in its own right, but also often highly conducive to improving agricultural productivity, food security, and nutrition. Gender-blind programs too often fail in these objectives and may even make matters worse by encumbering women with additional uncompensated duties or depriving them of control of resources in a manner detrimental to their welfare.

The renewed commitment to gender-responsive and gender-transformative agriculture now needs to be translated into more systematic policy actions, including integrating gender in agricultural strategies and programs. Research and development of agricultural technologies and interventions should begin with an understanding of how men's and women's interests as producers and consumers dovetail and diverge and should work to address the needs of women as well as men. Paying specific attention to closing the gender gap in assets is also expected to have long-run effects in empowering women and improving productivity and food security. Delivering on this potential will require systematic collection of sex-disaggregated data to improve our understanding of what works and a commitment to making actors at all levels, from government agencies to project staff, accountable for results. ■

EMPLOYMENT IN AGRICULTURE

Jobs for Africa's Youth

Karen Brooks, Sergiy Zorya, and Amy Gautam

AFRICA SOUTH OF THE SAHARA HAS THE WORLD'S FASTEST-growing population and the youngest. By 2050 the subcontinent, with its projected 1.7 billion people, will be the second most populous region in the world, after South Asia, and the only region in which the rural population will still be growing. Between 2010 and 2050 other regions will experience a significant decrease in rural population (which will fall by 50 percent in East Asia, 45 percent in Europe, and 10 percent in South Asia), while Africa south of the Sahara will add an estimated 150 million people in rural areas (an increase of nearly 30 percent). The young people yet to be born are in addition to the 330 million already present and about to enter the labor force, of whom 195 million live in rural areas.[1] As Figure 1 shows, the number of people entering rural labor markets each year is projected to increase until at least 2035.

Young people bring energy, vitality, and innovation into the workforce.[2] When their willingness to contribute is matched with opportunity, they can have a transformative impact on economic growth and social development. This impact is often referred to as the "youth dividend." African leaders know that the youth dividend will not be deposited automatically into national accounts; they will have to take proactive steps to collect it, and most are ready to do so. The past year has provided graphic evidence of the twin needs for agricultural growth and jobs. Yet again, global food prices have spiked; since June 2010, higher food prices have pushed nearly 44 million people into poverty.[3] Opportunities to create jobs and simultaneously lower food prices have been the subject of a number of recent events. For example, the "Young People, Farming, and Food" conference in Accra in March 2012[4] and the Fourth Conference of the African Union Ministers in Charge of Youth, held in Addis Ababa in September 2012,[5] focused specifically on these topics. At a special

Karen Brooks is director of the CGIAR Research Program on Policies, Institutions and Markets of the International Food Policy Research Institute in Washington, DC. **Sergiy Zorya** is a senior economist in the Agriculture and Environmental Services Department of the World Bank in Washington, DC. **Amy Gautam** is a freelance consultant who lives in Washington, DC.

FIGURE 1 Rural population share and number of people entering rural and urban labor markets annually in Africa south of the Sahara, 1950–2050

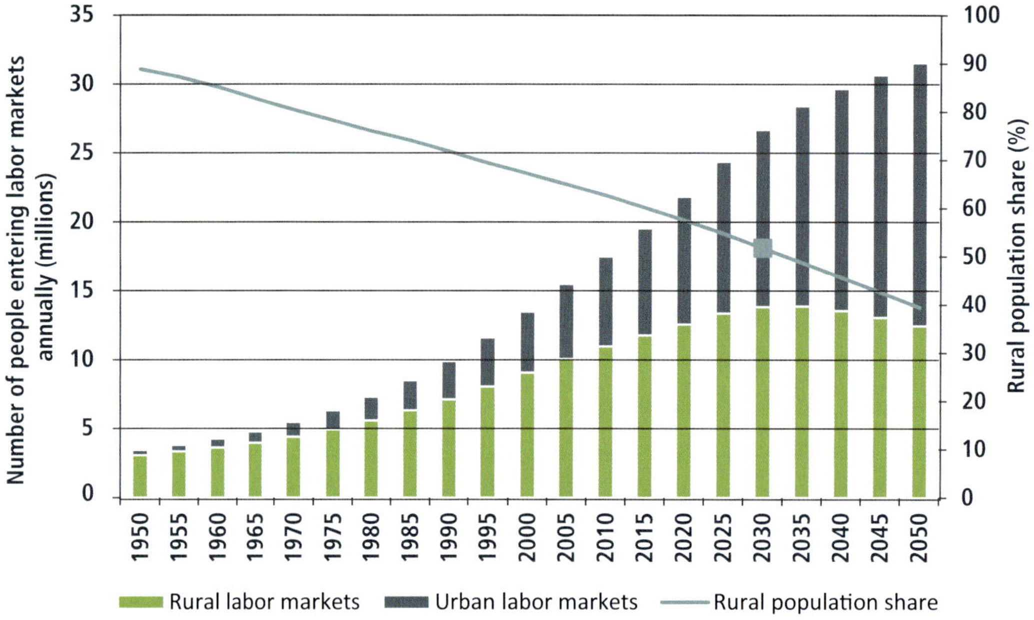

Source: B. Losch, S. Fréguin-Gresh, and E. T. White, *Structural Transformation and Rural Change Revisited: Challenges for Late Developing Countries in a Globalizing World*, Africa Development Forum Series (Washington, DC: World Bank, 2012).

session on youth in agriculture during the 2012 Farmers' Forum of the International Fund for Agricultural Development (IFAD), IFAD's president "emphasized the need to invest in the rural youth of today, the farmers of tomorrow."[6] Further, the United Nations' April 2012 Annual Ministerial Review recognized rural employment as the key to reducing poverty and food insecurity.[7]

Despite this recent attention, many may still perceive the topic of youth employment only in terms of formal jobs in the urban wage sector. Efforts to accelerate agricultural growth and improve food security are separated conceptually from efforts to create jobs for young people. This is a damaging compartmentalization, and if continued will likely result in the forfeit of Africa's youth dividend. Efforts to enhance agricultural growth and those to create employment for young people are complementary, and must be so understood.

BUT AS AFRICA DEVELOPS, WON'T AGRICULTURE DECLINE AND LABOR LEAVE RURAL AREAS?

Agriculture contributes one-quarter to one-third of African gross domestic product (GDP) but employs 65–75 percent of the labor force.[8] These data roughly replicate the historic experience elsewhere. A gap in labor productivity between the nonfarming and farming sectors on the order of two to one ordinarily draws people out of agriculture and into other sectors, such as manufacturing and services. Young people are leaving Africa's farms in large numbers. Forty percent of Africa's population already lives in cities, and urbanization is proceeding apace. But this surface similarity with historical experience belies a fundamental difference between Africa's experience now and that of elsewhere and earlier.

Africa's structural transformation (that is, the growth in incomes and resulting diversification of the economy) is taking place at a time of high global food prices and growing recognition of Africa as a reservoir for growth in agricultural production. Elsewhere and in the past, workers left farms for factories producing goods for local consumption; factory goods are now produced largely in Asia, and this will change only gradually. But the number of young Africans that can be absorbed into jobs in manufacturing and services even under optimistic assumptions is likely to be much less than the cohort of people now entering the labor force in rural areas.[9] Agriculture will likely continue to be the dominant sector of employment for most young people over the next few decades.[10] Even if all who can leave agriculture do so, high birth rates create a constant and growing pool of young people who will apply their energies and talents on the farmsteads and in the villages of their birth.

Fortunately, the high demand for agricultural products regionally and globally creates good opportunities for them to do so. Significant incremental growth can be achieved by improving the productivity of both land and labor without displacing labor. Average yields (and thus land productivity) in Africa are low compared with average yields in other regions and with estimated potential yields in Africa; increases can generate economic growth for many years to come.[11] Better management practices can increase resilience and reduce risk. Land can be brought into cultivation without encroaching on forests, and modest increases in mechanization can increase labor productivity. This is fundamentally different from the historical experience elsewhere.[12] With greater availability of tools and machinery, households now limited to working holdings of 1 to 2 hectares could manage more. With improved seed, fertilizer, plant protection agents, and timely advice, yields and profitability can increase enough to allow households to reduce their area devoted to staples consumed at home and expand into more lucrative crops and livestock products. And with higher profitability, the machinery and additional purchased inputs become affordable. Thus the availability of underutilized land in many places and substantial gaps in yields relative to their potential offer opportunities for agriculture to simultaneously raise labor productivity and absorb additional workers.

OPPORTUNITY IS REAL, BUT NOT YET RECOGNIZED BY YOUNG PEOPLE

According to the World Bank, the value of food on domestic markets in Africa is projected to increase from US$313 billion in 2010 to US$1 trillion in 2030.[13] World prices are high and expected to remain so for at least the medium term. Income growth, population growth, and urbanization in Africa are increasing the demand for imported food faster than the supply of domestically produced substitutes; food imports surged ahead of exports as recently as 2003 and continue to climb. Denser patterns of settlement are causing marketing costs to fall and returns to investments in primary processing of raw products to rise. This is the opportunity before African producers.

Local producers can capture thriving domestic and regional markets only if they become more competitive. Measures that reduce the costs of production (such as dissemination of improved technology) and the costs of marketing (such as investment in transportation and infrastructure) will enable increased profitability and reduce food costs. Even in countries that are relatively well linked to world markets, increased local production can bring down food prices because of friction in the transmission of international prices into local markets.[14] Improved agricultural productivity will thus benefit both producers and consumers (many of whom spend half or more of their budgets on food).[15] Lower food prices offer a secondary benefit by tempering demands for higher wages in the nonfarm sector, thus attracting new investment in manufacturing and services. New investment creates new jobs, fueling a virtuous cycle.

Opportunities on Africa's farms abound, but they are still unrecognized by or inaccessible to most young people. Agricultural employment is often not seen as a viable career, especially by young people in villages who see the reality of their families' situation. For example, for some countries, available statistics suggest that the typical farm is usually no larger than 1 to 2 hectares, and the most common implements are still the short hoe and machete.[16] In recent focus group discussions conducted for the World Bank, young, rural Africans

India's New Deal: Public Works and Rural Jobs

P. K. JOSHI

The Mahatma Gandhi National Rural Employment Guarantee Act (MGNREGA) in India—one of the largest and most ambitious social security and public works programs in the world—guarantees participating adults in rural households 100 days of paid manual labor each financial year. The government program was initially launched on a limited scale in 2006, a time of serious agrarian distress in India, to achieve inclusive growth by generating employment and reducing poverty in rural areas.[1] Its main objectives are to (1) provide livelihood security to the rural poor through guaranteed wage employment, (2) rejuvenate the natural resource base, (3) create a productive rural asset base, (4) stimulate the local economy, and (5) increase women's empowerment. A series of studies evaluating the program was published in 2012, and the findings demonstrate its success.

Since the program began in 2006, the government has spent a massive amount of money—about US$22 billion—on laborers' wages to remunerate the approximately 12 billion person-days of employment generated in rural areas. Since 2008, the public works program has existed in every district in the country and provides 50 million households with employment annually. In 2011–2012, the budget reached nearly US$8 billion.

The evaluation of the program, which itself is a result of the coordinated efforts of several Indian ministries to create productive assets in rural areas, found that nearly 866 million projects have been completed since 2006 and more than 1 billion are still underway. Public works projects include village roads, water conservation and harvesting systems, irrigation canals, renovation of traditional water bodies, and land development.

Most notably, the employment program provides many jobs to women: about 47 percent of people who received employment under the program were women, far exceeding the targeted 33 percent. Other major benefits include increased rural wages, reduced distress migration, barren and uncultivated land use, and empowerment of the weaker section of society. In some Indian states, microlevel evidence shows that groundwater levels have risen as a result of the program's public works.

Among beneficiaries, the program contributed to the consumption of more and better food, which leads to food and nutrition security, especially in underprivileged households.[2] Participants also spent more on nonfood items, and this growing demand is expected to trigger growth in the manufacturing and service sectors of rural areas.

There are three main reasons for this success of the guaranteed employment act: (1) strong government commitment, (2) proper targeting of the rural poor and underprivileged, and (3) employment opportunities during the lean season. To ensure its sustainability in the long run, the Indian government needs to converge and link this flagship program with rural development programs and other social protection programs, which would reduce spending without compromising efficiency, food and nutrition security, or progress in rural infrastructure.

P. K. Joshi is the director for South Asia at the International Food Policy Research Institute in New Delhi.

asked about the best and worst ways to earn a living rarely mentioned agriculture as a "best job," although they did not consider it to be the "worst." Good jobs are those that command good pay and respect, features not typically associated with farming in Africa now.

Young Africans already work on farms; 89 percent of rural young people are employed by their families or are self-employed, as are 71 percent of their urban peers, according to a recent large-scale World Bank survey of 10 African countries. And young people on farms are poorly educated—just over 30 percent have completed primary school.[17]

This quick sketch lays out the policy challenge for African leaders. Agriculture represents the sector of most immediate opportunity to realize gains in growth and to create employment for young people. The farming that can accomplish this must shift rapidly from low productivity and status to technical dynamism with recognized opportunity. The labor force that can best implement this transition is one that knows traditional agriculture and is young and full of aspirations, but this same group is poorly educated. What can African leaders do?

NOT OUR GRANDMOTHERS' FARMS...

Four basic employment options exist in agriculture for young entrants to the labor market, and each requires a different mix of capital, land, and skills (Table 1).[18] Policymakers seeking to accelerate job creation while raising productivity will need to adjust current agricultural programs so that large numbers of young people can transition into the most viable option for their circumstance. These options are compatible with those described by Felicity Proctor and Valerio Lucchesi in *Small-Scale Farming and Youth in an Era of Rapid Rural Change*. The authors note that 51 percent of households in a 2008 survey of nine countries in Africa south of the Sahara (SSA) reported that inheriting land already under cultivation was the most common means for their young people to obtain land, while 16 percent would be allocated land not previously cultivated, 9 percent would rent or borrow land, and 12 percent would buy land.[19]

OPTION 1. As already noted, more than half of young people will likely remain on the holdings of their families. They will need skills to manage higher-valued agriculture and some capital, but they already have land. These families could adopt a corporate approach to management of the household as an enterprise, as the additional skills and labor of multiple young adults in the household allow specialization. A combination of pooled off-farm earnings, a shift to higher-valued and more commercial products, and aggregation of household labor at peak periods could allow small farms to absorb young adults constructively. Many of these individuals could benefit from taking part-time off-farm work (eventually transitioning to option 3).

OPTION 2. A second group of young people will leave their childhood farms and establish new holdings, ideally larger than the parcels they left. These farmers have the greatest need for land, capital, and advisory oversight, and the highest potential for increased productivity. Few young farmers will be able to assemble the elements required to establish a new farm without assistance. New holdings may be in the localities where the young people already live and on land newly available for cultivation through clarification of ownership, conversion of marginal or grazing land, or public investment in irrigation and improvement. Alternatively, new holdings could be farther away, in which case establishment of the new farmsteads would require relocation. Resettlement is often controversial. Experience in Africa and around the world attests to the importance of strict adherence to voluntary decisionmaking on the part of participants, careful selection, full information for all stakeholders, effective support services for the new arrivals, and adequate investment in infrastructure. An assessment of several decades of public support for resettlement in Indonesia shows mixed results tending toward the negative. Improvements in settlers' incomes and access to public services were offset by disappointing outcomes in agricultural production, environmental degradation, and resentment against the newcomers by indigenous inhabitants.[20] Preliminary results regarding a program of market-assisted land reform in Malawi, in contrast, indicate that it may have had more positive outcomes.[21] The Government of Ethiopia has operated a substantial program of voluntary resettlement for a targeted group of vulnerable residents in densely settled parts of the country. The program has generated much attention and controversy, but a rigorous assessment is not yet available. If local young people can secure access to land in or near their communities, this is clearly the simpler approach. If relocation is required, lessons of past experience should be fully weighed.

OPTION 3. Higher-valued agriculture will use services more intensively, creating employment opportunities. Demand for transport, plant protection, veterinary services, mechanized field operations, and advice can be met by young people with skills and enough capital (or leased machinery) to start small businesses. Many of the individuals in this

TABLE 1 Options and requirements for agricultural employment

Employment option	Need for land	Need for capital	Need for skills
Option 1: Full-time on existing family holding	None	Medium	Medium
Option 2: Full-time on new holding	High	High	High
Option 3: Part-time combined with household enterprise	Low	Medium	High
Option 4: Off-farm wage work	None	None	Medium/high

Source: Authors.

third group may be part-time farmers, with small allotments of land and enough capital and skills to establish themselves as sellers of services.

OPTION 4. A fourth group will take formal or informal wage work on large commercial farms, in the processing and service sectors, or seasonally on small farms during peak periods. These young people need skills to handle a range of tasks and equipment. Wage work varies from low-skilled and low-paid day labor to jobs in primary production and processing that require and reward higher skills. Drivers, machine operators, mechanics, quality-testing technicians, and others will be in increased demand, and these jobs are better paid than unskilled day labor.

FINANCIAL SERVICES FOR MORE DYNAMIC AGRICULTURE

Access to capital and credit for smallholders has been a perennial problem, and the problem is magnified for young people. Various innovations to overcome the barriers and achieve sustainable outreach to smallholders have been developed, but the problem is far from solved, and risks and costs remain high.

Self-help groups, savings and credit associations, community-based financial organizations, rural and community banks, and cooperatives are among the organizational innovations developed to address the high costs and risks. Allowing alternative forms of collateral (such as chattel mortgages, warehouse receipts, and future harvests) can reduce the cost of credit. Leasing can be particularly helpful for young farmers, as it opens access to equipment with less collateral than is the case for purchase.[22] Linking agricultural credit to extension services can address the needs of young farmers for simultaneous finance and information.[23] Similarly, bundling credit with insurance may force farmers to purchase more insurance than they would otherwise wish, but at the same time opens new opportunities for borrowing. Matching grants are used in many programs supported by governments and nongovernmental organizations to promote improved technologies, empower farmers to hire service providers, build linkages with private firms through productive partnerships, and provide rural infrastructure for common use.[24] Many of the innovations in rural finance discussed here and below are still being tested, and their performance and sustainability on a large scale are not yet known. They nonetheless warrant close attention in the future, as successes are identified for replication and scaling up, while failed designs are avoided.

Outgrower arrangements can offer prefinancing of inputs and assured marketing channels. For example, contract farming with financial arrangements that limit risk of default or side-selling has proven effective in Mozambique, Rwanda, Tanzania, and Zambia.[25] Similarly, provision of credit through supply chains in Kenya has benefited more than 3,000 farmers. The program reduces the risk of default through cashless bank transfers.[26] The Alliance for a Green Revolution in Africa's Innovative Finance Initiative has achieved low default rates on partial guarantees in Kenya, Mozambique, and Tanzania. Rabobank's Rabo Sustainable Agriculture Guarantee Fund issues partial credit guarantees and other financial products to mitigate financial intermediaries' risks.[27] Even without partial guarantees, banks are sometimes interested in providing services to farmers if they have help training their staff to work with these clients.[28] Technology that brings banking services to rural areas is spreading quickly. Kenya's M-Pesa service has transformed rural banking there, and young people are especially quick to adopt innovations based on mobile phones.[29] Biometrics is being explored in the context of credit markets to protect banks from serial defaulters.[30]

Programs that combine access to financial services with advice or mentoring are likely to be especially suitable in light of the limited experience of young people. Those that start by offering grants and systematically transition to offering savings and ultimately credit may be promising. Young people should be integrated into the financial system rather than sequestered; young participants will eventually cease to be young but will probably remain farmers. They need to enter their middle years with good linkages to sound financial institutions and knowledge of how to use their products.

LAND FOR THE YOUNG

Population growth puts pressure on land. Many African societies are hobbled by ambiguities in transactability of land through purchase, sale, leasing, inheritance, assignment under traditional rules, and mortgage.[31] The fact that many young people are trapped on diminishing holdings while their governments seek to attract outside investors argues for urgent attention to easing constraints to transfer of land. Africa has room to accommodate its young people, especially if land allocations are accompanied by public investment in infrastructure to allow more profitable farming and simultaneously attract new ideas and investment from abroad.

Only about 10 percent of occupied land in SSA is formally registered.[32] Many countries have made recent progress in formally documenting individual landownership. For example, by the end of 2012, Rwanda had demarcated all 10.5 million land parcels in the country and registered and prepared leases for at least 83 percent of them. Ethiopia has awarded certificates for more than 25 million parcels in its rural areas. Madagascar has issued 75,000 certificates akin to traditional land titles, while in Tanzania about 27,000 certificates of customary rights of occupancy have been issued in two districts. Benin, Burkina Faso, and Côte d'Ivoire have piloted rural land use plans as another way to establish individual land rights.[33] Not all attention to registration need be for individual holdings. Registration of communal land can be an important step to improve security of tenure, particularly if an agreement with an outside investor or allocation of communal land to young people for new farm starts is contemplated.

African agriculture can absorb large numbers of new job seekers and offer meaningful work with public and private benefits.

Young entrants to the labor force, even those with little need for additional land, may benefit from improved security of tenure, the ability to engage in transactions, and relaxation of controls on rental. The very poor, the landless, the young, and migrants usually obtain land through rental if they do so at all, and constraints on rental transactions can hurt them in particular. Short-term rental or long-term leasing can facilitate labor mobility and can transfer land to more productive users. Rental can be particularly helpful in easing the intergenerational transfer of land; at present, elderly people hold most of the land even though in many cases younger family members or neighbors might be able to farm it better. The most common restrictions on rental markets, such as ceilings on rental rates or prohibitions against absentee landownership, are often introduced in an effort to safeguard the interests of smallholders. They may instead lock land into inefficient patterns of use and can greatly disadvantage young potential users.[34] In addition to regulatory barriers, the absence of an effective land tax allows those who hold land for speculative purposes to do so at little cost. An effective system of land taxation coupled with active rental markets could put idle land to work, often in the hands of young people.

Governments could take additional measures to open land to their young people. The extent of state landownership in SSA is largely unknown, as most lands have not yet been surveyed and registered.[35] Underutilized state land, once properly identified and surveyed, can be auctioned to the private sector to combine large-scale operators and small and medium farmers in innovative ways (although care must be taken not to disenfranchise indigenous users such as herders and subsistence communities). Special efforts can be made to accommodate young people who show promise in farming. Such programs would be most useful for people pursuing option 2, although benefits can be envisaged for others as well. Several programs of redistribution have been tried with a mixed record of success. South Africa's program has fallen short of its ambitious targets, probably in large part because access to land was not coupled with improved advice or mentoring and because a prohibition on subdivision meant that farmers could not secure the land in holdings of the size that they preferred.[36] Malawi's distribution of underutilized land from former tea estates seems to have met with greater success.[37] New pilot efforts are underway in Senegal and Zambia and should provide useful lessons.

BETTER EDUCATION AND ENHANCED MODERN SKILLS

Modern agriculture requires a wide range of specific skills. The rapid spread of mobile phones, and other information and communication technologies more generally, alters the nature of the skill set required, but it does not remove the handicap that incomplete primary schooling represents. To equip young Africans for success in the four options for agricultural employment, schools must do a better job of teaching them basic skills for any endeavor. Foremost among these are reading, writing, numeracy, and the ability to use digital technology and to access and interpret information.

In addition, a growing and diversifying agricultural sector will create nonfarm employment opportunities. Existing agricultural vocational schools can play a constructive role in training skilled personnel for jobs in processing, marketing, machinery operation and repair, transport and logistics, and quality control.

New agricultural extension programs can empower farmers to specify the information they require and to select the extension provider.[38] In these programs, information is recognized as a public good, with the government assuming a share of the cost, particularly for small farmers and the poor. Participatory and group-based approaches to extension (such as farmer field schools and farmer study circles) are gaining ground, and many of these models have application outside of the agricultural sector. While still imperfect in their execution, these newer extension models are likely an improvement over earlier, supply-driven models.

Other specific training and learning opportunities currently exist in many forms, such as competence-based training, local agribusiness development services, business incubators, rural productive alliances, and public-private partnerships.[39] Like some of the finance mechanisms discussed earlier, many of these programs are small in scale and experimental and have not been rigorously evaluated or tested for their effectiveness or sustainability on a larger scale.

The key objective of public policy should be to facilitate the access of rural people to information rather than specifying a rigid new agricultural curriculum. As dynamism in the agricultural sector picks up, returns to investments in skills will grow, providing people with incentives to acquire needed skills. An education sector that delivers quality at the basic level coupled with demand-driven agricultural advisory services and flexible short-term training will best serve the evolving needs. To foster such a system, Africa will need to reinvest in agricultural higher education. How best to do so under severe resource constraints requires careful thought and planning, which are already underway at the regional level.

STRATEGIES ALREADY EXIST AND SHOULD BE ENHANCED, NOT REPLACED

African leaders recognize the renewed importance of agriculture, and most are devoting resources to it. The African Union and the New Partnership for

Africa's Development have helped by developing the Comprehensive Africa Agriculture Development Programme (CAADP).[40] CAADP provides a conceptual framework to guide countries in their investment planning and a mechanism to share experience and mobilize support. The CAADP agenda centers on four key areas: land and water, access to markets, vulnerability and risk, and agricultural technology. None of these addresses an agenda uniquely for young people, but they all include specific issues that are particularly relevant to young people's needs. For example, in the area of land administration, as already noted, young people will benefit from policies that make rental and intergenerational transfer easier. In the area of agricultural technology, programs of mentorship and voluntary groups of people working with the same technologies can provide useful advice that compensates for young people's lack of experience. CAADP's experience-sharing mechanism allows countries to evaluate ways of relieving constraints to land, capital, and skills that hold young people back. CAADP can also use its advocacy work to help African leaders see how the agendas of agricultural growth and youth employment complement one another.

A number of initiatives point to the attention already being paid to the issue of youth employment: the Millennium Development Goals' Target 1.B; the African Union's African Youth Decade 2009–2018; the UN General Assembly's call for member states to prepare a "National Review and Action Plan on Youth Employment"; and the recent joint proposal among the African Union, the UN Economic Commission for Africa, the African Development Bank, and the International Labour Organization for a Joint Initiative on Job Creation for Youth in Africa.[41] Efforts to create opportunities for young people in agriculture thus do not require a new strategy. Attention should instead be trained on getting governments and their development partners to deliver on existing commitments, implementing investment programs, and making continued progress on policy and regulatory reform. Modest adjustments to meet the specific needs of the large group of young people can capture the complementarities of the agricultural and youth agendas.

CONCLUSIONS

Agriculture in Africa is now recognized as a source of growth and an instrument for improved food security. It is not yet seen as the major employer of the continent's young people, although in fact it is. This oversight must be remedied if Africa's leaders are to be able to design and implement the measures needed for the sector to deliver its potential. African agriculture can absorb large numbers of new job seekers and offer meaningful work with public and private benefits. For this to happen, constraints to land, capital, and skills must be addressed, and features to make programs friendly to the needs of the young enhanced. Existing programs in finance, land, and education and extension can be adjusted to address the specific needs of rural young people. The agriculture that attracts young people will have to be profitable, competitive, and dynamic. These are the same characteristics required for agriculture to deliver growth, to improve food security, and to preserve a fragile natural environment. Agriculture offers a handsome youth dividend that will benefit young people themselves, their wider societies, and the global community. With clarity of vision and political commitment from Africa's leaders, that dividend can be collected and widely shared. ■

Chapter 6

US AND EU FARM POLICIES

The Subsidy Habit

Jean-Christophe Bureau, David Laborde, and David Orden

WORLD AGRICULTURAL PRODUCTION EXPERIENCED ANOTHER tumultuous year in 2012. After a promising spring planting season, the United States suffered a severe drought, which pushed nominal prices of maize and soybeans in world markets to record levels. Some other large suppliers (Argentina, Brazil, China) increased their production of these two crops, but given the importance of the United States as a producer and the continued growth in demand, these increases were not enough to hold prices down. As a consequence, instead of declining from their peaks of 2008 and 2010/2011, feedgrain and oilseed prices remained high for a fifth consecutive year.

The agricultural policy response in developed countries has been bifurcated. The United States and other donor countries have increased their investments in overseas agricultural development projects and scientific research through initiatives such as the US Agency for International Development's Feed the Future program. Neither the United States nor the European Union (EU), however, has done much to expand productivity-increasing public agricultural research at home. The US Midwest and Great Plains, which helped feed Europe after World War II and have historically been a breadbasket for the world, are now responding to increased demand from biofuel production, which puts upward pressure on agricultural prices. The United States, the largest single producer of maize and soybeans, has at least refrained from limiting its food and feedstock exports, a welcome outcome that reflects a constitutional ban against export taxes[1] and the adverse consequences of earlier export embargos. Likewise, the EU has for the most part avoided using border policies to insulate the domestic market from rising world prices. This constraint on border policies has helped stabilize world markets.

Jean-Christophe Bureau is a professor at AgroParisTech in Paris, France. **David Laborde** is a senior research fellow in the Markets, Trade, and Institutions Division of the International Food Policy Research Institute (IFPRI) in Washington, DC. **David Orden** is a senior research fellow in IFPRI's Markets, Trade, and Institutions Division.

A Brazilian View of EU and US Agricultural Policy Reforms

ANDRÉ MELONI NASSAR

The US farm bill and the EU Common Agricultural Policy (CAP) have received surprisingly little attention in Brazil this year. In contrast to previous years when these policies were under review—especially 2001–2002 when prices were low—the Brazilian government, trade associations, and even the media have not given the issue due attention. Brazilian stakeholders believe that (1) food prices will remain high, which partially neutralizes the trade-distorting consequences of the US farm bill and (2) demand for agricultural commodities (including biofuels) will continue to grow rapidly, which diverts attention from the still-high trade barriers and export subsidies used in the European Union (EU). While government officials have expressed concern about the US farm bill, Brazil's private sector has not added the ongoing policy revisions in the United States and the EU to its agenda.

Much of the lack of interest in Brazil can be explained by the dormancy of the Doha Round, which essentially means US and EU legislators are not bound by any international obligations regarding the agricultural support given to their farmers; rather they are limited only by their own budgets. Without a resolution to the Doha Round, Brazilian stakeholders have only the World Trade Organization Dispute Settlement Body to act as a disciplinarian when policies between third-party countries are violated. Not only does the settlement process take a long time, but the settlements themselves have little or no capacity to constrain policy decisions made by national governments. Therefore, Brazilian stakeholders have no incentive to follow the revision process underway in the United States and the EU.

In general, the private sector is driven by short-term results, and experience with Brazilian trade associations shows that they too have limited interest in working on issues that have concrete outcomes only in the long run, such as the revisions to the US farm bill. The problems that Brazil's policymakers have with CAP are related to trade barriers and export subsidies, not with focusing on short-term issues, however. In the absence of EU export subsidies, the better strategy for Brazil is to pursue bilateral negotiations with the EU instead of waiting for multilateral solutions like Doha. On the other hand, the trade-distorting component of the US farm bill continues to be strong, and distorting payments only need lower prices to become harmful to the world market again, as they were in 1998–2002. The United States has been losing world market share, which reduces the negative effects of its policies on trade, but it is concerning from a Brazilian perspective that past Dispute Settlement Body findings are not being considered during the farm bill revision process. But in the event of lower world food prices and the absence of an agreement on the Doha Round, the Dispute Settlement Body is the only available way to resolve trade issues. Brazil will not hesitate to call on it if necessary.

André Meloni Nassar is the managing director of the Institute for International Trade Negotiations in São Paulo, Brazil.

In the high-price environment since 2008, one might expect that old policies of subsidizing farmers through price and income support would fade from the policy debate. But that is not the case. Both the United States and the EU are reviewing agricultural support policies and devising new options for supporting and protecting farmers, particularly in light of uncertainty about whether prices will remain high and volatile. Some of the resulting US and EU policy developments do not bode well for achieving the long-run policy goal of using undistorted markets to achieve global food security at the least cost. Assessing the effects of various countries' trade-distorting support and protection measures is often complex. Impacts are diverse among different groups of countries (exporters and importers), among different segments of their populations (net food producers and consumers), and under changing market conditions. In this intricate context, undistorted world markets, complemented with appropriate investments in a growing food supply and availability of social safety nets for those at nutritional risk, should be the centerpiece of the global food system.

By excessively protecting their own farmers from the risks of agricultural production, however, US and EU policies reduce the incentives for the world to cope with country-specific risk through a fair, efficient, and undistorted trade regime. They favor unilateral and noncooperative strategies instead of a multilateral, rule-based approach.

Movement toward the objective of undistorted markets suffered a blow with the lapse of the World Trade Organization (WTO) Doha Round negotiations after 2008. These negotiations, intended to build on the earlier 1994 WTO Agreement on Agriculture, aimed to reduce trade distortions by progressively reducing distortionary agricultural support and protection while giving countries latitude to support farmers in nondistorting ways. These could include government services, food security programs, and income support that is decoupled from production decisions (these acceptable support measures fall into the category known as the "green box").

In the absence of a new WTO agreement, constraints on distortionary agricultural support remain lax. The new US farm bill under debate in 2012 may well be a casualty of the failure of the Doha Round. Its programs may make it harder in the future for the United States to agree to support reductions such as those envisioned but not locked in by the Doha negotiations. Likewise, the EU is planning to continue high subsidies and is considering distortionary new support options.

US and EU agricultural policy developments in 2012 will have direct effects on markets and indirect effects that set the tone for agricultural support and border policies worldwide.

UNITED STATES: PROVIDING NEW ASSURANCES TO FARMERS

The 2008 US farm bill expired on September 30, 2012, and subsequently the post-election Congress extended it through September 2013. Thus, debate over the farm bill will continue. Still, the likely direction of US policy was discernible in separate bills passed in July 2012 by the full Senate and the House of Representatives Agriculture Committee. Under these bills, the United States would eliminate annual fixed direct payments made to farmers since 1996 and enact enhanced price or revenue protection that is more closely tied to production of specific crops. Total farm subsidy costs are anticipated to decline by 10 percent over the coming decade, but the new programs could make large payments in years of low yields or market downturns. Anticipated total expenditures over 10 years under the new law compared with continuation of past farm programs are shown in Table 1.

The new US farm bill would expand crop insurance subsidies. Driven by higher per-unit subsidy rates and expanded eligibility incorporated in previous legislation, as well as higher nominal costs correlated with higher crop prices since 2008, insurance payments (measured as total indemnities less farmer-paid insurance premiums) already exceeded the US fixed direct payments by 2011 (US$5.6 billion versus US$5.0 billion). With the drought-related losses in 2012, insurance payments will be even higher.

The new US farm bill would expand on the traditional insurance programs in two ways.[2] First, new subsidies are designed to make payments for specific crops when revenues for that crop decline below a targeted level within a production year. This new program is commonly called a protection against "shallow losses" that are less than the losses that trigger payments from individual farm

TABLE 1 Estimated effects of new US farm bill on expenditure levels, 2013–2022

	Billions of US$ (10-year totals)	
Farm bill title	Estimated spending under 2008 farm bill	Range of anticipated increase or decrease
Farm support programs	152.8	-14.1 to -14.4
Commodity programs	62.9	-19.4 to -23.6
Crop insurance	89.8	5.0 to 9.5
Conservation	65.2	-6.1 to -6.4
Nutrition	771.8	-4.0 to -16.1
All other titles	4.8	1.7 to 3.2
Total	994.6	-22.5 to -33.7

Source: Authors' summary based on US Congressional Budget Office estimates (March 2012 baseline and budget scores of farm bill proposals).
Note: Estimates are for fiscal years 2013 to 2022.

insurance. Although the drought of 2012 demonstrates the systemic weather-related risks associated with farming, and systemic risk provides one rationale for government intervention to address a market failure, existing within-year US crop insurance subsidies are already high.[3] Adding new insurance against shallow losses adds to this imbalance.

Second, the new bill would strengthen protection of farmers against multiyear losses for supported crops (feedgrains, soybeans, wheat, rice, cotton, and peanuts), such as when prices decline for several consecutive years. Two approaches dominated the debate on this part of the bill. A traditional approach would raise the nominal values of target prices that trigger payments to farmers when market prices fall below these fixed levels. The target prices set in the 2008 farm bill have been so far below market prices that they have offered essentially no farm support. The new higher levels would create a fixed price floor much more likely to generate payments in downturns from recent price levels. Such fixed price floors have a long tradition in US farm policy.

In the competing approach, the triggering mechanism for payments to farmers is not a fixed price level but a moving average of past revenue. Farmers would receive payments if revenue in a year dropped below some percentage of the moving average (proposals in 2012 were in the range of 75–89 percent). Proponents of this approach argue that it has a built-in policy design benefit.[4] Because the revenue support trigger moves with the market, if prices decline and stay at lower levels for several years, the level of revenue support will also move down. In contrast to the more traditional approach, farmers, while protected against too sharp an initial year-to-year revenue decline, would have to adjust over time to the lower revenue levels. However, initiating a moving average revenue program after a period of high prices, and particularly the very high prices of 2012, means that farmers would receive protection against the first revenue decline that might occur—for example, if prices come down from 2012 levels in 2013 or 2014.

Eliminating the fixed direct payments and strengthening price or revenue support based on current production of specific crops runs counter to efforts to reduce trade-distorting subsidies through the WTO. The fixed direct payments arguably fall within the WTO green box of programs agreed to have "no, or at least minimal, trade-distorting effects or effects on production."[5] In contrast, US crop insurance subsidies and the new shallow-loss and multiyear-loss protection programs are subject to an annual nominal support limit of US$19.1 billion. In the Doha negotiations that faltered in December 2008, a limit as low as US$7.6 billion was under discussion.

In the absence of a Doha agreement, US farm policy is evolving in a direction counter to the long-run objective agreed to multilaterally in the WTO. The new support programs will make it harder for the United States to agree to tighter constraints that may be negotiated in the future.[6] And other countries may feel little motivation to limit their own distortionary support or protection if the United States adopts a new farm bill along the lines that seem likely.

EUROPEAN UNION: CAP CONTINUITY AND MODIFICATIONS

The EU Common Agricultural Policy (CAP) is subject to the renewal of the seven-year Multiannual Financial Framework for the 2013–2020 period. This renewal process also provides an opportunity to revise the CAP. The Council of the European Union, incorporating the agriculture ministers of each member state, and the European Parliament, which share responsibility for determining the CAP, are currently discussing the proposal tabled by the European Commission on October 11, 2011.

From the early 1960s to the early 1990s, the CAP was characterized by market intervention. Most of the EU agricultural sectors were subject to administratively set prices, and authorities had to purchase excess production when market prices were lower than these fixed levels. High levels of price support kept EU production growing, while technical change raised yields and lowered costs. As a result, in the 1980s managing government-held surpluses became a substantial problem. The EU subsidized the disposal of surpluses abroad, which led to both large budget expenditures and world

An African View of EU and US Agricultural Policy Reforms

KWADWO ASENSO-OKYERE

The agricultural policies of the United States and the European Union (EU), which were both under review in 2012, help agricultural development in Africa, but they also subsidize their own farmers and impose stringent requirements for imports. Such policies tend to keep African agricultural goods out of US and EU markets.

US AGRICULTURAL TRADE AND AID POLICIES

The US farm bill gives policymakers an opportunity to assist US farmers, but it also has implications for farmers and consumers in developing countries, especially in the absence of an agreement on the Doha Round of multilateral negotiations. It provides commodity price support and other payments to US farmers that tend to suppress world prices and distort market conditions. Such situations serve as disincentives for African producers.

The US policies on biofuels—which cover assistance for the production, marketing, and processing of biofuel feedstocks—reduce the amount of cereals and pulses on the world market and raise their prices. In turn, African countries pay higher prices for their food imports, which ultimately raises concerns for food security in Africa. On the flip side, the US Food for Peace program provides food aid to many African countries in an effort to reduce vulnerability in food insecure households. The program was very helpful during the droughts and accompanying famines that struck the Horn of Africa and the Sahel in 2011–2012 and during previous years when high prices made food inaccessible to many people in Africa.

EU AGRICULTURAL TRADE AND AID POLICIES

Although the EU has opened up its markets to African producers through the Cotonou Agreement and the Everything But Arms agreement, among others, the EU Common Agricultural Policy (CAP) makes it difficult for those producers to be competitive. The EU protects its markets by (1) keeping prices artificially low through subsidies for farmers, processors, and exporters of agricultural produce and (2) imposing nontariff barriers such as strict health and safety regulations (for example, restrictions on maximum residue levels). By subsidizing their own farmers at the current magnitude, European agricultural produce (primarily beef, poultry, and tomatoes) can be sold in Africa at prices so low that African producers cannot even compete in their own countries despite the advantage for low-cost production. The food safety regulations in the EU make it difficult for African countries, which typically rely on smallholder producers who use unsophisticated technologies, to export to Europe and can wipe out any benefits the producers gain from not having to pay tariffs. Such consequences have already been observed for horticultural produce from many African countries. For African farmers to take advantage of the EU's preferential treatment, they need better capacity to increase productivity and maintain food safety. The EU can provide assistance to make this possible.

Kwadwo Asenso-Okyere was a senior research fellow in the International Food Policy Research Institute's Eastern and Southern Africa Regional Office in Accra, Ghana.

market distortions that triggered international disputes and retaliation. In the 1990s the cost of storing commodities and subsidizing exports became so high that the EU engaged in reforms. Since then, continual reforms (in 1999, 2003, and 2008) have led to the progressive dismantling of the intervention system and the de facto end of export subsidies, as shown in Figure 1.

Farmers have been compensated for lower EU price supports by direct payments. In the 2000s these payments were decoupled from production to the extent that farmers are no longer required to produce crops or animals to receive support. The payments are considered to fall into the WTO's green box and thus are exempt from a nominal cap. However, the €42 billion handed out each year necessarily indirectly raises output by keeping some farmers in production (even though they do not need to produce to get support), easing credit constraints, and lowering risk aversion. The EU farm support policy is now based almost entirely on the direct payments, which have been made conditional on requirements regarding the environment, animal welfare, workers' safety, and other social regulations.[7]

The European Commission's proposal does not depart significantly from the CAP reform movement initiated 20 years ago but would remove most of what is left of market management. Because of the new institutional power of the European Parliament, which gained full joint decisionmaking power with the Council in 2010, the proposal also reflects the concerns of elected representatives and their farm constituents. In particular, ministers of agriculture and the Agriculture and Rural Development Committee of the Parliament expressed

A Chinese View of EU and US Agricultural Policy Reforms

FUNING ZHONG

Grain security is ensured when the quantity of total grain supplies can adequately satisfy national needs and when, domestically, grain is priced such that those who need it can afford as much as they need. It is always a top priority in Chinese agricultural policy. To fulfill demand, China has become a net importer of major cereal crops, bringing in an estimated 5.5 million metric tons of corn, 2.9 million metric tons of wheat, and 2 million metric tons of rice in the 2011/2012 crop year alone. In order for China to grow as much grain (including soybeans) domestically as it currently imports, it would need to increase the areas currently sown to grain crops by more than 30 percent. At the same time, Chinese consumers have been struggling with high food prices, which increased by nearly 7 percent in the first half of 2012—a spike that is higher than any seen throughout the entire previous year. As both grain imports and prices increase significantly in a seemingly irreversible trend, China is beginning to more carefully evaluate the farm policies of its major trading partners to assess how they impact food security within China's borders. Accordingly, China's farm policy may move either toward more openness to imports or away from it by placing greater emphasis on a self-sufficient domestic food supply.

In 2012, the United States proposed its new farm bill, which may partially shift budget support from commodity programs to agricultural insurance programs. This could help stabilize production and, to some extent, prices for insured crops. While the proposed changes to the existing US farm policy would likely have only a moderate impact on prices themselves, they could have a lot of influence on reducing the fluctuation of those prices. As such, the potential outcomes do not go against China's policy goals. The farm bill also proposes that fixed direct payments are replaced with price or revenue support based on current production of specific crops. China may be able to tolerate these changes, however, because the basis for becoming a net importer of crops is the shortage in domestic food supply, not the necessity to compete with lower prices.

The European Union's (EU's) 2012 proposal for a new Common Agricultural Policy (CAP), on the other hand, may have negative implications for China's grain security. A significant reduction in the CAP budget as well as the "green payment" stipulation (which states that farmers must meet certain requirements related to crop diversification, grasslands, and ecological focus areas before receiving direct payments) may lead to lower production levels of major EU crops. Although the EU is not a major source of imports for China, any significant reduction in production from such a major player in trade would inevitably lead to either a reduction in exports or an increase in imports. China also needs to keep a close watch on the greening trend and find ways to adapt to its outcomes. As more major trading partners take similar measures to reduce greenhouse gas emissions and protect the environment, the total food supply in the world market might be tightened as its prices increase. This will certainly put heavier pressure on China's grain security from a long-term perspective and could lead to the traditional self-sufficiency argument, which is currently regaining some support. In general, rather than taking further measures at its borders, China should enhance efforts to push and strengthen domestic production.

Funing Zhong is a professor in the Department of Agricultural Economics at Nanjing Agricultural University in the People's Republic of China.

FIGURE 1 Composition of the EU agricultural budget, annual expenditures, 1990–2010

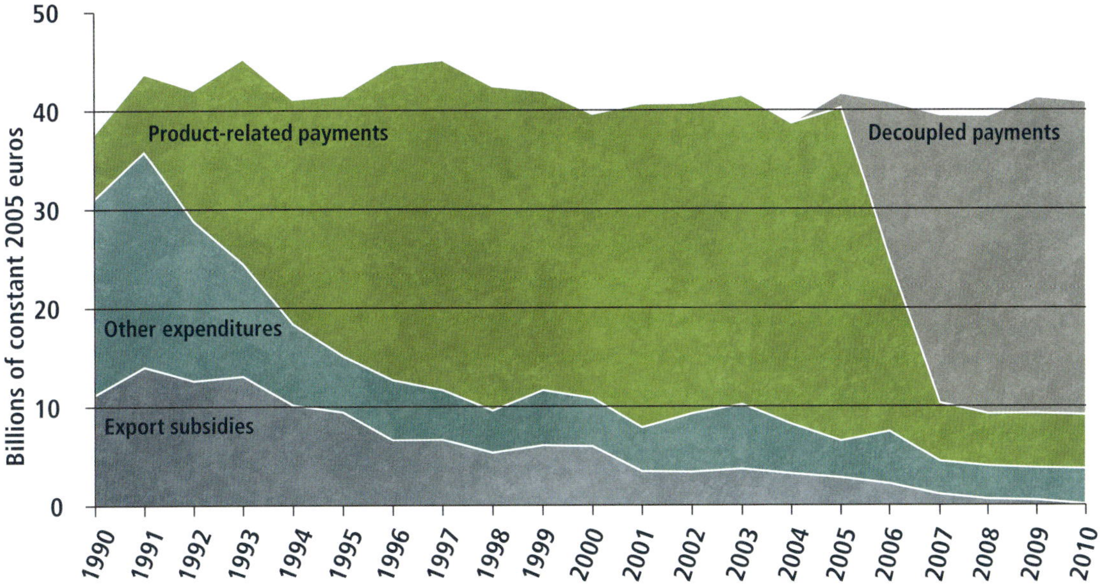

Source: J.-P. Butault and J.-C. Bureau, based on EU Commission budget data.

concerns about further market liberalization because of a possible increase in price volatility. They also raised concerns about leaving European farmers dependent on market forces while their US counterparts benefit from systems of insurance and countercyclical support and while major emerging-market countries are raising production subsidies.

These concerns are reflected in the Commission's proposal for coping with potential "crisis" periods of exceptionally low prices. The proposed crisis package includes a tendering process for some products (barley, maize, rice, and beef stocks) and private storage aid for others (sugar, olive oil, flax, beef, butter, skimmed milk powder, pig meat, and sheep meat). These measures would be funded from a €3.5 billion reserve separate from the CAP budget. The Commission's proposal also authorizes member states to develop national-level insurance and income stabilization tools with some cofinancing from the EU budget, but with ceilings that ensure that these new programs will remain limited.

The EU budget for direct payments will remain high under the CAP, since the Commission proposes maintaining this expenditure in nominal terms. National allocations ("envelopes") for direct payments would be adjusted so that those receiving less than 90 percent of the EU average payment per hectare would receive more, moving all EU member states toward more uniform payments per hectare by 2019. The largest gap would be reduced by one-third.

The most controversial issue is the Commission's proposal to reorient the direct payments, with an increased requirement for environmental measures.[8] A basic payment scheme would replace the current single farm payment scheme. Under the new design, the basic direct payment would continue to be subject to relatively minimal requirements. An additional payment of 30 percent of the total would be conditional on farmers complying with three measures: (1) crop diversification (farmers would have to cultivate at least three crops a year on the land they do not set aside); (2) an "ecological focus" requiring that farmers devote 7 percent of their land to a conservation area where biodiversity is protected; and (3) maintenance of permanent pastures. The ecological focus is particularly opposed by farmers' organizations.

In addition to the management of markets and the support of farm income, the EU has over time developed a rural development component of the CAP, sometimes called the second pillar, which is cofinanced by member states. The Commission has proposed some changes in this policy, with new priorities. The rural development measures would include compulsory funding for climate change mitigation and adaptation and new land management measures, including organic farming.

There are many reasons for dissatisfaction with the Commission's proposal. Some fundamental inconsistencies of the current CAP persist. For example, maintaining basic direct payments means also maintaining the undesirable effects of the current system, in particular the capitalization of payments into land prices and the push toward specialization of farms. The Commission has introduced new payments to promote crop rotation and to help young farmers overcome barriers to entry, while the problems these measures address are actually caused or at least worsened by the system of direct payments itself. There is little left of the idea of reallocating support toward public goods, which was extensively discussed during the preparation of the proposal. And making 30 percent of the direct payments conditional on specified farm management requirements (for example, crop rotation with an ecological focus) is a high-cost policy compared with targeting payments directly to the provision of public goods, such as water management, carbon storage, or biodiversity preservation.[9]

The Commission's proposal does not change the basic direction of the CAP or address its most fundamental problems. It continues to shift support to direct payments and rejects attempts by parliamentarians and farmers' organizations to turn these payments into risk-based and countercyclical instruments, or to implement a large-scale, EU-wide farm insurance system. While many observers lament the proposed reform's lack of ambition, it has managed to keep most of the bad ideas proposed for CAP reform out of the agenda. Overall, the reform should not result in further market distortions.

Although 20 years of reforms have not lowered EU farm support, they have shifted the support to payment categories that have so far been exempted from WTO-imposed reductions. While a new crisis package may, in a period of market collapse, call for export subsidies to support prices, it is clearly a policy of last resort in the proposed configuration. Even if a Doha agreement is eventually completed, there is no reason for the EU to withdraw from its 2004 commitment to end export subsidies (which it currently does not use). The EU's main concern about a Doha agreement concerns the prospect of increasing access to the European market and the need to lower tariffs, which is largely independent from the CAP reform. A sharp reduction in EU agricultural tariffs, in particular in the dairy, beef, and sheep sectors, would likely lead to large imports and hurt EU cow-calf producers and the extensive farming sectors that are still central to the rural economy of some European regions. Incomes are very low in these economically fragile sectors. It is unlikely that the EU will be willing to endanger so many farmers without substantial concessions from other countries.

DIVERGENT POLICIES BUT CONTINUED US AND EU SUPPORT

There are few areas of convergence between the 2012 US farm bill proposals and the European Commission proposal for 2013 CAP modifications. Both proposals share concerns about price and income volatility, but the EU responds to them with much more limited instruments than the existing and proposed US arsenal of measures. The new US farm bill is likely to enact increased protection of farmers against possible adverse events, moving it further from the spirit of previous multilateral negotiations, which aimed to reduce distortions in world markets. Meanwhile, EU policies will continue to rely largely on payments decoupled from direct links to production of specific crops and livestock. But its transfers to farmers will remain very high compared with the United States. The events and political momentum of 2012 perpetuate a global regime of support to farmers in the major developed countries.

Continuation of these support policies has detrimental direct and indirect effects on global food security. To the extent that these policies induce higher US or EU production and lower world

Grain Drain: Agricultural Policies in the Post-Soviet States

SERGEY KISELEV

Although there were no major changes made directly to the agricultural policies of the post-Soviet states in 2012, several external factors influenced agriculture in the region. For example, after becoming a member of the World Trade Organization in August, the Russian Federation increased subsidies for young breeder cattle purchases (to the equivalent of more than US$30 million) in an effort to enhance the competitiveness of its livestock. Similarly, the Russian government introduced milk subsidies to help this sensitive sector adapt to the liberalized market. Payments were initiated for hectares of cultivated land based on the volume of crop yields, bioclimatic potential, and soil fertility. Some tax concessions have also been prolonged, including the zero-profit tax for agricultural producers. More specific to 2012, given the agricultural damage resulting from negative climatic conditions in the spring and summer, the government developed mitigation measures for the 20 regions affected and allocated 6 billion rubles (about US$200 million) to compensate agricultural producers for their direct losses.

The Customs Union of Russia, Kazakhstan, and Belarus has also had an effect on agricultural policies in the region, especially as the union agrees upon and enforces its new technical regulations and health codes. Given Russia's World Trade Organization accession, the common market for these countries has high requirements to ensure that their agricultural products are competitive.

In Ukraine, milk, meat, and egg production increased in 2012 and managed to stabilize the livestock industry, but poor climate conditions led to decreased production of some crops, including grain, sugar beets, and sunflowers. Despite this reduced production, export levels were relatively high for major agricultural products such as cereals, oilseeds, and vegetable oil. In fact, Ukraine's total agricultural exports in 2012 exceeded the previous year's by 38 percent, and by the end of November 2012, total grain exports exceeded 11 million tons, with wheat contributing almost 47 percent of that total. This provoked concern about national food security and a discussion on the possible limitation of grain exports (by introducing quotas, administrative barriers, or bans) to refocus attention on domestic food needs while maintaining high rates of nongrain export crops.

Grain exports have also been relatively high in the Russian Federation and could exceed 12 million tons by the end of the year. The high volume of exports has not been discussed in Russia, however, because the country has carryover grain stocks to ensure national food security and restricting exports at present would have potential negative consequences.

It was also a difficult year for Kyrgyzstan, where grain yields were more than 50 percent lower than in 2011. Achieving food security there and in other Commonwealth republics, such as Moldova, will require an increase in either grain (mostly wheat) imports or food aid.

Sergey Kiselev is the director of the Eurasian Center for Food Security at Lomonosov Moscow State University, Russian Federation.

prices, and increase uncertainties for overseas producers, they reduce incentives for agricultural development elsewhere. They contribute to a concentration of world production in a limited number of countries, increasing the risk exposure of the global food system. For the United States, these distortionary effects will be larger if prices decline from recent high levels; that is, the distortions will be greater just as other producers face more adverse conditions. For the EU, distortions arise from the sheer amount of income transfers. In reaction to the continued agricultural subsidies in the United States and the EU, other countries will maintain protection and develop unilateral support programs for their own farmers, lessening hopes for achieving a world food system that effectively provides for global food security at the least cost. ■

Chapter 7

REGIONAL DEVELOPMENTS

Policy Choices on the Ground

APART FROM 2012'S GLOBAL CHALLENGES AND OPPORTUNITIES for food policy, important developments with potentially wide repercussions took place in individual regions and countries. This chapter offers perspectives on food policy developments in various regions, including growing efforts to improve resilience to famine in Africa, a slow reform of food subsidies in some Arab countries, political uncertainties surrounding modern food-retail systems and crop technologies in South Asia, the intricacies of rice production and trade in East Asia, and concerns in Latin America about how to manage the region's role as an agricultural powerhouse sustainably.

Africa

Ousmane Badiane, Kwadwo Asenso-Okyere,
Samuel Benin, and Tsitsi Makombe

In 2012, African countries continued their efforts to improve food security and agricultural growth under the Comprehensive Africa Agriculture Development Programme (CAADP). These efforts included new global and regional partnerships, such as the Global Alliance for Action for Drought Resilience and Growth, the Global Alliance for Resilience Initiative–Sahel (AGIR), the New Alliance for Food Security and Nutrition, and the Dublin Process to forge a science and technology partnership between CAADP and the CGIAR Consortium. Launched in April 2012, the Global Alliance for Action for Drought Resilience and Growth represents a joint commitment by African governments and international development partners to make building resilience and promoting economic growth in the Horn of Africa and the Sahel a central part of development efforts. Both the Global Alliance for Action for Drought Resilience and Growth and AGIR can be enhanced through policies to (1) raise agricultural productivity and incomes, (2) increase regional trade to stabilize domestic markets, and (3) expand and improve delivery of social services (education, health, and safety nets) so as to maximize their impact on vulnerable groups' resilience.

AGIR AND FOOD SECURITY IN WEST AFRICA AND THE SAHEL

While 2012 food production in the Sahel fell below the record level of 2010, spreading the fear of famine, production was still well above the 2006–2010 average. Moreover, when cereal imports were factored in, available supply exceeded demand in every country, even without food aid. Supply in the Sahel and West Africa as a whole exceeded demand by more than 0.6 million metric tons and 2 million metric tons, respectively.[1]

The Need for a Broad Resilience Framework

The Sahel region is therefore not faced with a sudden, short-term crisis, but rather with chronically vulnerable communities who even in "normal" production years may still be suffering. A positive by-product of the alarming predictions has been the launch of AGIR in June 2012, by regional organizations and their global partners, to restore and protect the livelihood base of vulnerable communities through social safety nets, nutrition programs, emergency food reserves, and enhancement of the communities' production capacity and assets. The initiative also seeks to strengthen regional food and nutrition security governance capacities. In all these areas addressed by AGIR, West Africa offers a host of best practices to other parts of Africa. These include

1. a regional Food Crisis Prevention and Management Charter, adopted in 2012, that defines the objectives, principles, and modalities that should guide responses to crises at the regional and country levels;
2. a harmonized framework (Cadre Harmonisé Bonifié, or CHB) that defines vulnerability indicators and procedures shared by multiple countries in the region for evaluating if the indicators point toward a crisis, initiating action, and monitoring progress; and
3. a governance and intervention coordination mechanism, the Food Crisis Prevention Network, which was created in 1984 and revamped in 2012.

The Role of an Emergency Food Reserve

A key component of the food crisis charter is the regional emergency food reserve, adopted in September 2012. The reserve of 0.4 million metric tons is to support country responses and is triggered when an evaluation based on CHB modalities indicates that a predefined vulnerability threshold is reached. The full cost of the reserve over eight years is estimated to be about US$1 billion. Funding is to come from a 0.5 percent tax on aggregate imports imposed under the "Zero Hunger in West Africa" initiative; this tax is expected to raise about US$3 billion over the same eight-year period.[2]

The Trade and Technology Dimensions of Building Resilience

Countries have, under the food crisis charter, committed to free regional trade in food products, but neither the charter nor the strategy for the reserve include practical steps for exploiting the potential contribution of regional trade to stabilizing domestic food supplies and mitigating the impact of production shocks. More important, by lowering local market instability and thus the need to use emergency food reserves, regional trade lowers emergency reserves' size and operational cost.

To effectively address vulnerability, countries have to boost agricultural productivity levels among smallholder communities. As shown in Figure 1, countries in the Horn of Africa and the Sahel are located in regions (East, Central, and West Africa) with the lowest levels of labor productivity. Reversing this requires fostering the pace of technological innovation through better policies and increased investments in research, irrigation, market infrastructure, and institutions. The emerging CAADP-CGIAR partnership, which seeks to align research priorities of the Consortium with the needs of national CAADP investment plans, could make a considerable contribution in this area and needs to be strengthened.

FIGURE 1 Land and labor productivity growth in Africa, 1980–2005

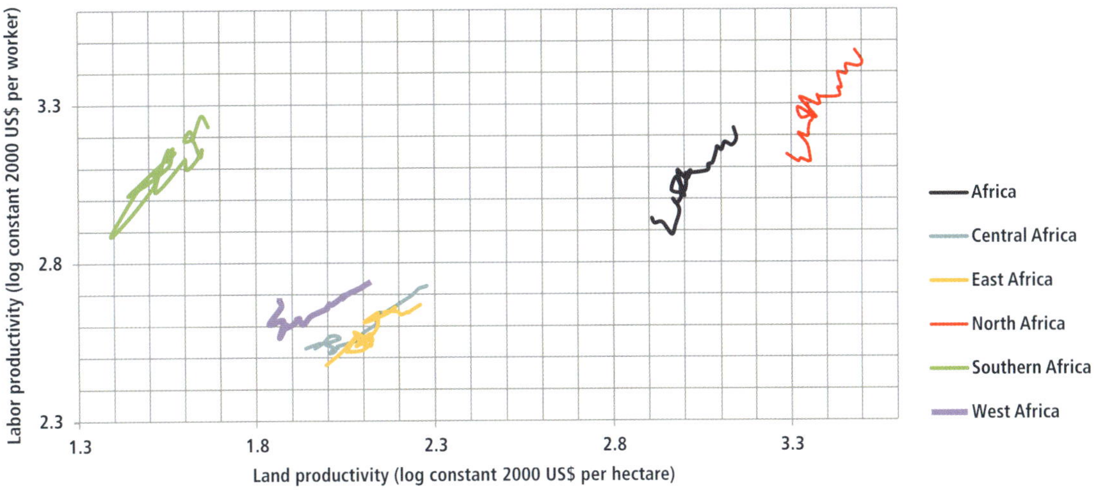

Source: S. Benin, A. Nin-Pratt, S. Wood, and Z. Guo, *Trends and Spatial Patterns in Agricultural Productivity in Africa: 1961–2010*, ReSAKSS Annual Trends and Outlook Report 2011 (Washington DC: International Food Policy Research Institute, 2012).

POLICY RENEWAL AND ITS IMPORTANCE FOR BUILDING RESILIENCE

Major efforts are being made by African nations and the global community to significantly increase investments in agriculture and improve the sector's policy environment. By 2012, 60 percent of African countries had signed a CAADP Compact and prepared sector-investment plans to boost agricultural growth. In 2012, the Economic Community of West African States established a regional agency to coordinate implementation of its regional investment plan, budgeted at nearly US$1 billion (on top of US$24 billion declared in investment plans by its 15 member countries).[3] In addition, more countries are making progress, albeit slowly, toward the CAADP target of a 10 percent agricultural sector budget.[4] Countries also continued the transition to evidence-based policy planning and implementation by initiating institutional and technical capacity needs assessments, leading to the establishment of 15 country Strategic Analysis and Knowledge Support Systems.

In response to efforts by African countries to improve food security, the G8 launched a New Alliance for Food Security and Nutrition during its 2012 summit. Under the Alliance, G8 members intend to work with African governments and local and international private-sector actors to lift 50 million Africans out of poverty over 10 years. The Alliance seeks to accelerate implementation of CAADP country-led plans by increasing private-sector investments and implementing the G8 countries' L'Aquila pledge to mobilize more than US$20 billion over three years in support of global food security. The Alliance was launched in Ethiopia, Ghana, and Tanzania and will expand to other countries. More than 45 multinational and African companies have pledged over US$3 billion in investments, while G8 countries are following up on their L'Aquila commitments: the Global Agriculture and Food Security Program, set up in response to the L'Aquila pledge, has awarded a total of US$430 million, US$160 million of which was awarded to 11 African countries in 2012.[5]

Policy Reversal Would Set Back Progress and Undermine Resilience

While the renewed emphasis on agriculture and increased investments in the sector are a welcome change, some of the accompanying policy actions may have serious drawbacks. For example, recent policy measures to introduce input subsidies

Malawi: Macroeconomics, Small Farmers, and Short Food Supplies

DYBORN CHIBONGA

The year 2012 saw some significant macroeconomic policy developments in Malawi that had implications for smallholder farmers, their families, and their communities. Malawi's government has prioritized agriculture and food security in strategic national policies such as the Malawi Growth and Development Strategy II, the national budget, and the Economic Recovery Plan. It has also adopted the Maputo Declaration, which calls on African governments to allocate at least 10 percent of their resources toward agriculture in order to realize 6 percent annual agricultural growth. Malawi has, in response to the Maputo Declaration, developed a harmonized agricultural investment plan called the Agriculture Sector Wide Approach. In addition, since 2005, the government has implemented the Farm Input Subsidy Programme, which has contributed to Malawi's becoming food secure at the national level.

Despite its history of supporting agriculture, however, the government's decision in May 2012 to devalue the national currency has aggravated current economic hardships. Consumer purchasing power has declined as a result of the devaluation, as well as subsequent depreciation and rising inflation and fuel costs. Such conditions have had far-reaching consequences for the cost of farm inputs such as seeds, fertilizers, herbicides, and labor. At the same time, financial institutions have increased their lending rates, making it virtually impossible for smallholder farmers to access financial services. Thus, the costs of doing farming business and purchasing food have increased tremendously.

In addition to these financial difficulties, some districts are experiencing acute food shortages because of climate change. According to a report by the Famine Early Warning Systems Network, which provides an update on food security conditions in Malawi, the number of people in need of humanitarian assistance between October 2012 and March 2013 will have jumped from 1.63 million to 1.76 million, but response plans by the government and development partners remain inadequate. Smallholder farmers feel the effects of climate change on a daily basis. It is therefore disappointing that the 18th session of the Conference of the Parties to the United Nations Framework Convention on Climate Change in Doha, Qatar, in late 2012 did not come up with a work program in agriculture as had been envisaged at the 2011 climate change conference in Durban, South Africa.

Among those trying to help smallholder farmers deal with these challenges and improve their situation is the National Smallholder Farmers' Association of Malawi (NASFAM). With a membership of 108,000, NASFAM is the largest farmer-led business and development organization in the country and operates in all three regions of Malawi. The members primarily grow food and cash crops. NASFAM promotes the farming business among its members and counts on an enabling policy environment for its business.

NASFAM is optimistic that 2013 will be a good year, in spite of recent challenges. Two promising developments are the adoption of conservation agriculture by NASFAM members and the appropriately huge investment in high-quality seeds for members. However, NASFAM is concerned that if the rains allow farmers to harvest a good crop, they might also increase rates of inflation, currently in double digits. NASFAM is hoping that the government's Economic Recovery Plan will help stabilize the economic situation.

Dyborn Chibonga is chief executive officer of the National Smallholder Farmers' Association of Malawi, Lilongwe.

present risks. Although the commitment by governments to act boldly must be saluted, the fiscal and economic costs of these programs are rapidly getting out of control. In Malawi, for example, fertilizer subsidies alone accounted for 74 percent of the Ministry of Agriculture's budget and nearly 7 percent of the country's GDP in 2008/2009.[6] This level of spending is undoubtedly unsustainable and threatens the very existence of the programs. Programs in Ghana and Zambia are also trending in that direction. In the spirit of CAADP, countries should prioritize policies and instruments that are fiscally sustainable, smallholder oriented, and private sector friendly.

Other recent efforts include the establishment of an Agricultural Transformation Agency in Ethiopia and an Agricultural Transformation Agenda in Nigeria. Both programs have broad mandates that go beyond policies to include investments, service delivery, institutional innovations, and capacity building. Such a comprehensive agenda is likely to have significant macroeconomic and welfare ramifications beyond the intended objectives. While designing these new initiatives, countries need to avoid a reversal of the reform process under CAADP and a return to the policies of the 1960s and 1970s that resulted in macroeconomic imbalances and poor agricultural performance and led to the painful and costly structural adjustment reforms. Countries should instead seize the opportunities for dialogue, review, and critical analysis offered by CAADP and forge effective policies to achieve their goals.

Better Management of Public Resources

African countries cannot find sufficient resources fast enough to meet both the social services needs of a large group of poor and vulnerable people and the investment needs of accelerated agricultural growth; hence, they need to look at social services provision from the point of view of maximizing their impact on labor productivity among poor households.

Figure 2 illustrates the scope for maximizing the productivity impact of social services, using results from a study of the effects of health expenditures on disease prevalence and on labor productivity among rural Ugandan households. It summarizes the efficiency impact of an increase in different categories of health expenditures by poor and nonpoor households. Across all households, expenditures on consultation services for all diseases have by far the largest impact in terms of reducing morbidity-related inefficiency. These findings suggest that health programs, in this particular case, can contribute more to raising agricultural labor productivity by emphasizing access to consultation services. Analysis of other programs in the health, education, and social protection sectors may yield similar results.

CONCLUSION

The underlying chronic vulnerability in the Horn of Africa and the Sahel requires broad and long-term action beyond the typical short-term, emergency

FIGURE 2 **Efficiency effects of a 10 percent increase in different types of household health expenditures, Uganda**

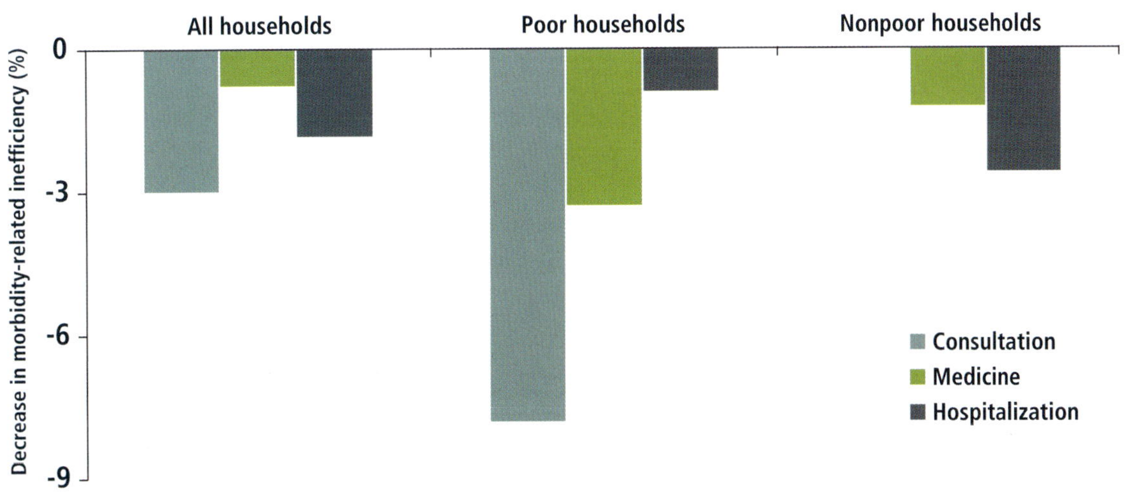

Source: O. Badiane and J. Ulimwengu, "Short-Term Social Needs and Long-Term Growth: The Productivity Impact of Social Expenditures," in *Social Protection in West Africa: The Status Quo, Lessons from Other Regions, Implications for Research*, Thematic Note 3 (Washington, DC: International Food Policy Research Institute, West and Central Africa Office, 2012).

response. West Africa has made significant progress in forging responses that represent best practices. Building resilience will also require continued commitment to the CAADP investment and policy-renewal agenda, including greater cross-border trade as well as improved management of public-sector expenditures, so as to maximize their impact on labor productivity among the poor and vulnerable.

Ousmane Badiane is director in Africa for the International Food Policy Research Institute (IFPRI), Washington, DC. **Kwadwo Asenso-Okyere** was a senior research fellow in IFPRI's Eastern and Southern Africa Office, Accra, Ghana. **Samuel Benin** is a research fellow in IFPRI's Development Strategy and Governance Division, Washington, DC. **Tsitsi Makombe** is a program manager in IFPRI's West and Central Africa Office, Washington, DC.

Arab World

Clemens Breisinger, Perrihan Al-Riffai, Olivier Ecker, and Nadim Khouri

The Arab revolutions of 2010–2011 brought dramatic changes to Egypt, Libya, Syria, Tunisia, and Yemen. In 2012 each of these countries—with the exception of Syria, which remains mired in civil war—took steps toward greater democracy by electing new presidents, parliaments, or both. History shows, however, that countries in transition are vulnerable to civil conflict, particularly if progress in accelerating job-creating growth and improving food insecurity is slow. Slow progress in these areas was, in fact, one of the triggers for the Arab Awakening, and governments in the region are now well aware of the importance of food security for political stability. This overview will mainly focus on those Arab countries in transition that have undergone or initiated political and socioeconomic reforms. Where comparisons are possible, the section will reflect on the rest of the region.

ECONOMIC TRANSITION AND FOOD SECURITY

Although other Arab countries such as Jordan and Morocco are also undertaking political reforms, the countries going through the most dramatic changes are Egypt, Libya, Tunisia, and Yemen. These countries have undergone rapid political and economic transitions and therefore face greater immediate challenges than more gradually reforming countries. The public debate in these four post-revolutionary countries shifted from a strong focus on political transition in 2011 to increasingly include economic matters, including food security, in 2012. This is partly because economic recovery after the revolutions has been slow. After sharp recessions in Libya, Tunisia, and Yemen, and modest growth of 1.8 percent in Egypt in 2011, economic growth picked up slowly. It was projected to reach 2.2 percent in Egypt and 2.4 percent in Tunisia and remain negative at -1.9 percent in Yemen in 2012.[7] The generally unfavorable global economic environment and uncertainty about political and economic reforms have contributed to the sluggish recovery, negatively affecting trade, investments, consumption, and inflows of development aid.

Food security remains a key development challenge at the national and household levels. Given the high food-import dependency of all postrevolutionary countries, higher global food prices led to rising food-import bills in 2012 despite higher domestic cereal production in Egypt, Tunisia, and Libya, compared with 2011.[8] At the same time, foreign-currency inflows from exports, tourism, remittances, and other sources to finance food imports have stagnated or declined, leading to an overall worsening of food security at national levels. The exception is Libya, where hydrocarbon output and exports have grown rapidly in 2012 and recovered to preconflict levels. While the region has been marked as "on track" to achieve most of the Millennium Development Goals, particularly in primary school enrollment and gender parity in primary education, these success stories are overshadowed by the deterioration in numbers of the undernourished. The region of West Asia and North Africa is the only world region in which the number of undernourished people has increased over the past 20 years, from 13 million in 1990–1992 to 25 million in 2010–2012.[9]

Perception-based surveys suggest that household food security has worsened in 2012 in postrevolutionary Arab countries and has improved in other oil-importing and oil-exporting Arab countries (Figure 3). Among the four postrevolutionary countries, the situation of households has deteriorated most dramatically in Egypt and Libya, followed by Yemen. Only in Tunisia has household food insecurity slightly improved since 2011. These trends may be explained by the generally slow economic recovery, which meant zero or even negative per capita income growth in Egypt, Tunisia, and Yemen. The fact that Libya's households have become more food insecure shows that factors other than economic growth and the purchasing power to import food, including the distribution of incomes and access to quality services, affect food security. Rising prices and depreciating currencies have also depressed real incomes and negatively affected nutrition. Extensive food subsidy schemes helped to buffer these effects and to maintain relatively modest food price inflation in Egypt and Tunisia, which likely prevented further increases in food insecurity in those countries.

SUBSIDIES AND THE "DOUBLE BURDEN OF MALNUTRITION"

To address this difficult situation in 2012, newly elected governments have primarily continued to rely on the same policy measures that most Arab countries implemented during the 2008 global food crisis and the 2010–2011 uprisings. Favored measures among postrevolutionary and other Arab countries were increases in public-sector wages and food subsidies, while fuel subsidies were largely unchanged in countries that have them; Morocco, Syria, and Tunisia raised their fuel prices in an attempt to reduce spending in the face of extreme fiscal challenges.[10] Food insecurity and poverty likely would have further increased without those measures. However, those measures also contributed to dramatically increasing the budget deficits in all oil-importing countries plus Yemen, thus jeopardizing the sustainability of the current fiscal situation. For example, in Egypt government subsidies for fuel and food make up 8 percent of GDP, and food subsidies alone make up slightly more than 2 percent of GDP.[11] Of those food subsidies, 60 percent are allocated to baladi bread, of which the average Egyptian household consumes between 17 and 22 loaves a day at a fixed price of £E0.05 per loaf.[12]

Food subsidies have been criticized for straining budgets, generally being poorly targeted, and often being misused (past misuses include exporting subsidized bread or feeding it to animals). What is less well known is that food subsidies may also contribute to simultaneous under- and overnutrition, a problem referred to as the "double burden of malnutrition." The double burden of malnutrition may occur not only within a nation or household but even within the same individual: for example, someone might suffer from both deficiencies in micronutrients and obesity. Among the Egyptian population chronic malnutrition is still widespread among preschool-age children, while a large share of the adult population is obese as a result of calorie-heavy diets. This phenomenon has often been linked to income inequality and the coexistence of poverty and affluence, as well as a lack of nutritional awareness and dietary knowledge. Both forms of malnutrition carry high economic costs because they reduce people's productivity and strain the healthcare system that must treat the disorders. Hence, Egypt's food-subsidy system should be improved so that the subsidies are fiscally sustainable and better targeted, encourage healthy food habits, and help reduce both under- and overnutrition.

The region of West Asia and North Africa is the only world region in which the number of undernourished people has increased over the past 20 years.

REFORMING GOVERNANCE

Inefficiencies of social protection instruments such as food and fuel subsidies are one of the many indicators that show a lack of governance in the region. A number of governance indicators show either stagnating or deteriorating governance performance in the Arab countries. Transition countries have experienced the worsening performance of two critical governance indicators: (1) corruption and (2) voice and accountability (the second measures perceptions of citizens' power to choose their government). The performance of these two indicators over the last decade is believed to have contributed to political and social unrest at the beginning of 2011.

In 2012, several Arab countries initiated long-term policies and investments, including governance changes. Tunisia started to improve incentives for private investments and initiated job-creation programs. In addition, the government of Tunisia established a new Ministry of Regional and Local Development. Algeria, Bahrain, Morocco, and Yemen increased spending on infrastructure and education. Several Gulf Cooperation Council (GCC) countries invested in increasing the volume of their strategic grain reserves. Saudi Arabia announced plans to set up a body to monitor foodstuff prices and the availability of food within the kingdom in an effort to improve food security. Morocco, Sudan, and Tunisia cautiously started to reform subsidies, and Egypt announced similar plans. Yemen began discussing the decisive implementation of its National Food Security Strategy. Moreover, civil society groups in countries such as Egypt, Jordan, and Lebanon took steps to establish charity organizations such as food banks.

FUTURE POLICY INITIATIVES

More strategic foresight and action would help achieve a food-secure Arab world. The "old" debate on the need for enhanced policies and investments in trade, agricultural production, or social protection policies is still ongoing. The Arab region has suffered from the worldwide trend of decreasing investments in agriculture. However, there is a danger that redressing this situation, a response that has received more attention recently, may well be seen as the "silver bullet" to solving food security issues in the region. The Doha Declaration of November 2012 calls for increasing spending on agriculture and food production, and the Qatar National Food Security Programme has announced large investments in greening the domestic desert. One key reason often identified by GCC countries and others for focusing on domestic production is the perceived risk that international food markets may cease to function, with potentially devastating effects on net food-importing countries. Given GCC's fiscal surpluses and sound finances, it would be desirable for these oil exporters to address their own concerns about international food markets by seeking more clarity on the specific costs and benefits of—and synergies between—agriculture and trade-based solutions at subregional and national levels. Non-GCC countries, being fiscally strained, will have more limited choices.

Given the high levels of malnutrition, particularly among children, more emphasis should be placed on food security at the household level, enhanced education, and health and nutrition policies. Inarguably, governance improvements in postrevolutionary countries, as well as many other Arab countries, will be instrumental in the successful implementation of policies and investments and thus long-term development in general.

This multifaceted approach to food security received an important boost by the Committee on World Food Security in 2012, with its creation of a Global Strategic Framework for Food Security and Nutrition.[13] This framework goes beyond the elements of agriculture, trade, and social protection in the ongoing debate to include issues such as excessive price volatility, smallholder-sensitive investments, nutrition, land tenure, gender concerns, and other issues.

Political will and commitments will be essential in the process. The League of Arab States is an important interlocutor in regional debates over future courses of action. Through the intergovernmental mechanisms of the League of Arab States, Arab decisionmakers have agreed on a regional strategy and regional emergency plan, which were

FIGURE 3 Perception of food insecurity in Arab countries, 2009–2012

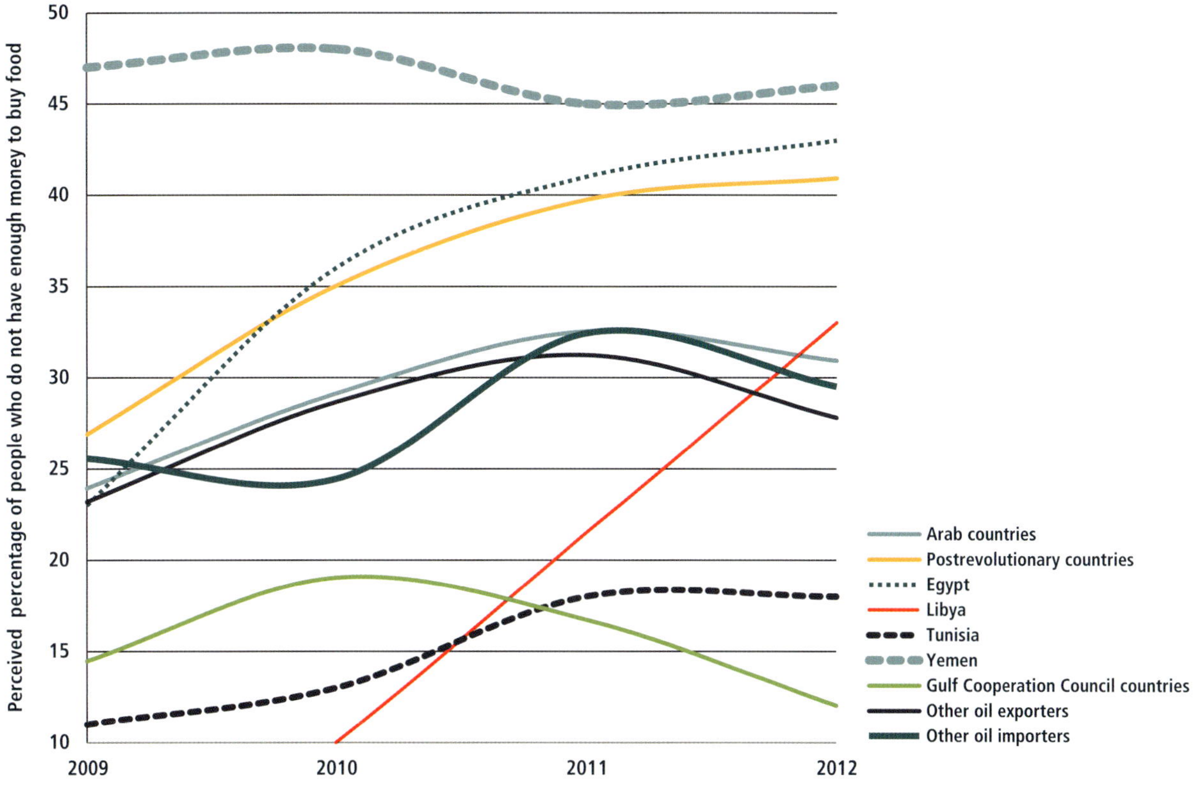

Source: Authors' representation based on data from Gallup World Poll, "WorldView," https://worldview.gallup.com/signin/login.aspx, accessed January 11, 2013.
Notes: Postrevolutionary countries include Egypt, Libya, Tunisia, and Yemen. Gulf Cooperation Council countries include Bahrain, Kuwait, Oman, Qatar, Saudi Arabia, and the United Arab Emirates. Other oil-exporting countries include Algeria, Iraq, and Sudan. Other oil-importing countries are Comoros, Djibouti, Jordan, Lebanon, Mauritania, Morocco, Somaliland (a region within Somalia), Syria, and West Bank and Gaza. Data for groups of countries are weighted by country population size. No estimates are available for Libya for 2009 and 2011. Estimate for Morocco is for 2008, not 2009.

negotiated and agreed upon in 2007. Could a thorough revision and revival of these documents be the starting point for remobilizing regional and national efforts in addressing food security?

Clemens Breisinger is a research fellow in the Development Strategy and Governance Division of the International Food Policy Research Institute (IFPRI), Washington, DC. **Perrihan Al-Riffai** is a senior research analyst in IFPRI's Development Strategy and Governance Division. **Olivier Ecker** is a research fellow in IFPRI's Development Strategy and Governance Division. **Nadim Khouri** is the deputy secretary of the United Nations Economic and Social Commission for Western Asia.

South Asia

P. K. Joshi and Devesh Roy

Record food-grain production in South Asia during 2011–2012 was followed first by late, scanty rains in India and Nepal and then floods in some parts of Bangladesh and northeast India during the 2012 rainy season (May–August). Food-grain production during the monsoon season is expected to suffer as crop-sown areas shrink in size and too much or too little water has a negative impact on yields. Overall, agriculture is performing poorly in South Asian countries compared with some Southeast and East Asian countries. South Asian agriculture's average annual growth during the last decade was a little more than 3 percent: this is roughly the same as in other Asian regions, but South Asian growth is not accelerating over time and is likely to remain around 3 percent. South Asian agriculture is constrained by a lack of appropriate policies for accelerating agricultural production, reducing food prices, and opening trade barriers. All South Asian countries are hampered by delays in reforming the agricultural sector. The key reasons cited for these delays are (1) policy paralysis due to political uncertainty; (2) policies without any economic logic that are intended to appease political constituencies—these tend to have high costs that obstruct agricultural reform; (3) weak governance structures, which lead to huge policy implementation gaps; (4) weak capacity to execute policies in effective and efficient ways; and (5) powerful bureaucracies, which control, regulate, and delay reforms. However, a few policy decisions have been made at both the country and regional level.

FERTILIZER SUBSIDIES IN NEPAL

Fertilizer access is important to South Asian agriculture. In Nepal, inadequate supply and delayed distribution of fertilizer (together with scanty rainfall) severely limited agricultural production, especially of rice and maize, during the 2012 rainy season. Because about 80 percent of Nepal's food grains are produced during the rainy season, any adverse effect on food production during this period seriously impacts the country's food security. The region consumes more than 1.5 times the fertilizer it produces, implying that a significant share of consumption must be filled through imports. Some of South Asia's fertilizer needs could be met within the region, however.

Rising fertilizer subsidies have a major influence on South Asian countries' fertilizer access. Subsidies keep Indian fertilizer cheap relative to Nepalese fertilizer: urea costs approximately US$96 per metric ton in India, but more than US$240 per metric ton in Nepal's local market. As long as Nepal provided no fertilizer subsidy of its own, an estimated 70 percent of the country's annual consumption of 350,000 metric tons of fertilizer was smuggled in from India.[14] Although Nepal has since adopted fertilizer subsidies, the price imbalance continues and threatens a loss of fertilizer business for Nepalese farmers.

To overcome such problems, Nepal's government has decided to nearly double the fertilizer subsidy for fiscal year 2013. The Ministry of Agriculture and Development is holding discussions to increase the subsidy, currently US$35 million, to approximately US$63 million, according to ministry sources. The subsidy increase should check smuggling and will allow the Agriculture Inputs Company Ltd., a government entity that imports and distributes fertilizer, to procure 300,000 metric tons from India for fiscal year 2013, double the current amount.

Thus, Indian price and subsidy policies influence fertilizer prices and use in Nepal, as well as in Bangladesh. Given fertilizer policy's importance, regional policymaking needs to take into account how to benefit small farmers in South Asia.

FOREIGN DIRECT INVESTMENT IN MULTI-BRAND RETAIL IN INDIA

In India, modern food retail has been estimated to have grown 49 percent annually from 2001 to 2010.[15] Foreign direct investment (FDI) will further accelerate modern food retail's growth. The argument in favor of FDI in multi-brand retail, which includes food commodities, is that multinational retail chains will bring not only capital investments to India, but also institutional best practices and knowledge on how to coordinate the procurement, distribution, and sale of a large number of products that modern supermarkets sell.[16] Promoting FDI moved forward in 2012, when the Indian government made the long-awaited decision to approve FDI in multi-brand retail, with a cap of 51 percent on how much investment can come from foreign firms. This decision will have implications for agricultural marketing, especially improving the food value chain. Agriculture and agricultural marketing are controlled by the states, and many state governments had reservations about allowing FDI in multi-brand retail in their states. Multinational companies must therefore seek state-government permission for multi-brand retail investments.

The following are a few benefits for the food system expected from FDI in multi-brand retail: (1) reduction in postharvest losses and waste; (2) improvement in marketing efficiency; (3) reduction in retail overheads; (4) assurance of food quality and food safety; and (5) provision of more employment opportunities in ensuring high-quality standards through such operations as cleaning, physical grading, and packaging. In developing countries, consumers pay three to four times the farmgate prices for fresh produce. About 60–80 percent of the price consumers pay goes to commission agents and wholesalers to cover transportation, loading, unloading, storage, reducing waste that occurs between the farm and retail outlets (especially of perishable commodities such as fruits, vegetables, milk, or meat), overheads, and profits. In organized retail markets, however, consumers pay only one-and-a-half to two times the farmgate prices for basic food. Further, integrating smallholders with modern supermarkets allows farmers to sell their produce directly to modern retailers, cut transaction costs, and reduce market risks, thereby making a greater profit than under the old arrangement. Thus, more efficient markets will be a win-win proposition for farmers and consumers. Allocating 50 percent of FDI to back-end processes has the added advantage of improving production and minimizing losses. However, the Farmers' Forum, apprehensive that a large part of the agricultural produce in modern retail outlets may be imported, proposed that retail outlets be required to procure a minimum of 75 percent of their agricultural products from Indian farmers.[17]

SOUTH ASIAN SEED BANK

Quality seed is a prerequisite for increasing agricultural production, but most farmers in South Asia are deprived of such seed. The seed replacement rate—that is, the percentage of seed planted in a given season that was freshly acquired by farmers rather than saved from an earlier harvest—is too low (ranging from 5 to 15 percent). Recognizing the importance of seed, the South Asian Association for Regional Cooperation (SAARC), which includes eight South Asian countries, agreed in late 2011, at its 17th summit, to establish the SAARC Seed Bank to help farmers get quality seeds in case of a shortage due to natural calamities. The objectives of the Seed Bank are to (1) provide regional support to national seed security efforts, (2) address regional seed shortages through collective action, and (3) foster intercountry partnerships. This initiative involves creating a regional seed security reserve for the member states and requires each country to contribute at least 1 percent of its total seed requirements to the bank's reserve. The seed contributed will subsequently be available to South Asian countries during uncertain situations such as crop failures resulting from floods or drought.

The Seed Bank will help member countries make quality seeds available. The initiative is expected to facilitate adoption of improved varieties; increase the seed replacement rate; and help member countries share among themselves best practices, technologies, and techniques for producing quality seeds.

Besides establishing the Seed Bank, participating heads of state at the SAARC summit renewed their commitment to alleviate poverty and reduce income inequalities within their societies. They reaffirmed their resolve to improve their people's well-being and quality of life through people-centered sustainable development. They also recognized the importance of fully implementing the South Asia Free Trade Agreement to move toward the creation of an enabling regional economic environment. The leaders also resolved to address the SAARC Seed Bank's operational issues by the next Council of Ministers session in 2013 to ensure the bank's effective functioning.

BRICS SUMMIT IN INDIA

India organized the Fourth BRICS (Brazil, Russia, India, China, and South Africa) Summit in New Delhi on March 29, 2012, with the overarching theme of "BRICS Partnership for Global Stability, Security and Prosperity." The summit's purpose was to give further momentum to the BRICS

Vision 2021: Bangladesh Charts a Path toward Food Security

MUHAMMAD ABDUR RAZZAQUE

Motivated by a strong political commitment, Bangladesh has taken a comprehensive and coordinated multisectoral approach to the pursuit of food security. The country's policies for ensuring food security are guided by its Vision 2021—a program of policies and actions culminating in the year of Bangladesh's golden jubilee—and its national development goals, as well as the Millennium Development Goals. Among other things, the Vision 2021 seeks to ensure that by 2021 most people have a minimum of 2,122 kilocalories per person per day and that the incidence of poverty is lowered to 15 percent. In addition, the 2007–2008 global food crisis was instrumental in shifting policy toward food self-sufficiency.

Agriculture plays an important role in the country's food security strategy. In 2011–2012 the government continued to promote agriculture by, for example, ensuring easy availability of high-quality seeds and fertilizers; maintaining reduced prices for non-urea fertilizer and a targeted subsidy for diesel fuel; expanding irrigation; developing new high-yielding varieties, including stress-tolerant and short-duration varieties; providing easy access to credit; ensuring remunerative prices for farmers; and allocating funds for research. In June 2012 Bangladesh produced an all-time record of 33.9 million metric tons of rice—more than three times the harvest of 1971, the year of the country's independence. Positive results were also recorded for wheat, vegetables, and potatoes, and the livestock and fishery sectors continued to thrive, bringing agricultural growth to more than 4 percent in recent years. Despite the government's efforts, however, production centers heavily on rice, and substantial deficits still prevail in the production of pulses, meat, eggs, milk, fruits, and vegetables. To help diversify the food basket at the household level, the government has undertaken the program "One Household, One Farm."

In response to more frequent shocks from natural disasters and more volatile food-grain prices, the government has scaled up public stocks and distribution of food grains. The country's effective storage capacity was increased from 1.4 million metric tons in 2009 to 1.8 million metric tons in 2012 and is projected to reach 2.1 million metric tons by the end of 2014. This increased storage capacity should help extend the shelf life of grain, reduce leakage, and improve the country's capacity to respond to shocks.

Bangladesh has also strengthened its safety net. The budget allocation to social safety net programs has risen in recent years, with nearly 25 percent of households covered by some type of safety net program in 2010.[1] Public distribution of food grains rose from 1.3 million tons in 2007–2008 to 2.3 million tons in 2010–2011, largely to help consumers cope with the spike in the price of food grains. Distribution through open market sales and fair price cards proved effective in this regard. When prices fell in 2011–2012, public distribution was reduced. In addition, the 100-Day Employment Generation Program, which employs the extremely poor, accounts for an estimated 4.2 million person-months of employment during lean seasons and has helped eliminate seasonal food insecurity from northwestern Bangladesh. The government's efforts have not only helped reduce food inflation (below 3 percent in September 2012), but also facilitated an increase in the rice-wage equivalent. Overall, the incidence of poverty fell from 40 percent in 2005 to 31.5 percent in 2010–2011, and access to food improved. Nevertheless, the prevalence of malnutrition among children and women is still of paramount concern in Bangladesh.

In 2011, in view of the food security challenges, the government revised the Country Investment Plan, which consists of 12 high-priority programs for improving availability of, access to, and utilization of food. The plan involves investments totaling US$7.8 billion of which US$5.1 billion is awaiting funding, including first-priority investments worth US$3.4 billion.

The country's recent actions have amply reflected its commitment to food security. It is important now to sustain growth and diversify production, drawing on technological innovations and agricultural research, especially to develop stress-tolerant crop varieties. There is a need to expand the knowledge base on agriculture and food security, and continued support from development partners will be crucial. Finally, Bangladesh must enhance its safety net programs and develop effective community-based nutrition programs.

Muhammad Abdur Razzaque is the minister of Food and Disaster Management and a member of Parliament in Dhaka, Bangladesh.

process. The association of BRICS countries is a unique development that allows these countries to address common opportunities and challenges.

The summit's Delhi Declaration confirmed the BRICS countries' commitment to address various economic and political issues of global and regional importance. Country leaders also released "The BRICS Report," which focuses on synergies and complementarities among the BRICS economies and highlights the economies' role as drivers of world economic growth. An updated edition of the BRICS Joint Statistical Publication was also issued at the summit. Government ministers attending also agreed to adopt the "Action Plan 2012 to 2016" in order to achieve the following objectives:

- Creating a basic agricultural information exchange system among BRICS countries (coordinated by China)
- Developing a general strategy for ensuring the most vulnerable populations' access to food (coordinated by Brazil)
- Reducing the negative impacts of climate change on food security and adapting agriculture to climate change (coordinated by South Africa)
- Enhancing agricultural technology cooperation and innovation (coordinated by India)
- Promoting trade and investment (coordinated by Russia)

South-South cooperation has now become an important part of reducing poverty and ensuring food security.[18] There are several examples of partnerships that demonstrate the critical role of collaboration among Brazil, China, India, and South Africa in driving innovations (such as knowledge sharing, capacity building, and improved technologies) to reduce world poverty and hunger. The summit reaffirmed the commitment of all BRICS countries to learn from each other to meet global challenges and harness emerging opportunities.

MORATORIUM ON FIELD TRIALS FOR GENETICALLY MODIFIED CROPS IN INDIA

Although some policies were reformed, food policy also suffered a significant setback in 2012. In an unfortunate development, the Supreme Court–appointed Technical Expert Committee in India recommended, in October 2012, a 10-year moratorium on field trials of any genetically modified (GM) crops. The controversy arose after Bt eggplant, a GM crop, was barred from being released for cultivation on the pretext of concern for food safety, biodiversity, and the environment. The committee's recommendations go one step beyond prohibiting GM crops' release for cultivation and do not allow even field trials of any new GM crop varieties. The recommendation is in contrast to scientific institutions' goal of developing transgenic products to overcome biotic and abiotic stresses in cases where conventional breeding's probability of success is too low.

Several GM products have been developed for this purpose, including maize, rice, rapeseed, potato, tomato, and cotton. Bt cotton was released and approved for large-scale production in India and has completely transformed Indian cotton production: cotton-lint production increased from 1.6 million metric tons in 2000 to 5.7 million metric tons in 2012, and exports went up from merely 127,000 metric tons to 1.5 million metric tons during the same period. Also, Bt cotton has reduced production costs by reducing the need to use pesticides, which were used indiscriminately to control bollworms. Nevertheless, Bt eggplant was not approved for release for cultivation. Eggplant is largely grown by smallholders in poverty-ridden areas who could have benefited immensely from reduced pesticide costs and increased production. Such decisions, which are not based on scientific logic, will have negative effects on frontier research and demand-driven technology generation.

P. K. Joshi is director in South Asia for the International Food Policy Research Institute (IFPRI), New Delhi. **Devesh Roy** is a research fellow in IFPRI's Markets, Trade, and Institutions Division, New Delhi.

East Asia

Kevin Z. Chen

Despite a much-anticipated slowdown in 2012, East Asia still enjoyed the highest economic growth rates of any world region.[19] Many East Asian countries are undergoing rapid economic transformations involving improved living standards and decreased poverty. How best to manage these transformations and address threats associated with rapid development, such as income inequality, environmental degradation, and the sustainability of future growth, is at the center of policy debate.

In order to ensure sustainable future growth, the region needs to prioritize balanced growth strategies that include both export-oriented and domestic-demand-driven growth. East Asia also needs to diversify its production bases and sell more value-added products.[20] Reducing inequality between urban and rural residents has been addressed by increases in health and education spending, and improved social protection programs. And because agricultural productivity growth is an important part of economic transformation, rural and agricultural infrastructure and farmer access to inputs and extension services should be promoted.[21]

Another important part of ensuring regional food security is providing adequate and stable rice production. Unlike in 2007–2008, rice is in ample supply in the region, and China increasingly relies on cheap rice imports from Vietnam, Cambodia, and Pakistan, probably because of Chinese rice production's declining comparative advantage as a result of rising rural wages.[22] Reliance on imports of other products is more risky, however: while the rice price has been stable, the 2012/2013 prices of US-grown maize, soybeans, and wheat are forecast to rise 22 percent, 19 percent, and 12 percent, respectively, as a result of drought.[23]

This presents a problem for China, Indonesia, and other East Asian nations that currently import huge quantities of grains and soybeans from the United States (Indonesia temporarily halted soybean import taxes and expanded the portfolio of commodities controlled by the state purchasing agency, Bulog, to include soy).[24] The vast majority of these grain imports go into animal feed. Asia, which accounts for more than one-third of the world's maize imports, and especially China, which is the world's second-largest maize consumer as well as its largest soy consumer, would suffer considerably from continued dry weather in the United States.[25]

It is particularly worth noting that, according to the recent release of trade data from China Customs, China imported 2.1 million metric tons of rice, 3.6 million metric tons of wheat, and 5.0 million metric tons of maize in the first 11 months of 2012 after many years of near self-sufficiency in those grains. This may signal that China has reached a turning point in terms of its grain security.

The goal of adequate domestic food production has prompted a number of significant country-specific policy initiatives. China's 2012 No. 1 Document focused heavily on agricultural research and development, which China has prioritized since 2000 and in which it is outpacing both Brazil and India and catching up to the United States in terms of public investment levels.[26] Myanmar has continued to open up, and reforms, including a new agricultural bank that provides loans to farmers and other agriculture-related clients and support for seed-production companies, are under way.[27] These reforms may pave the way for Myanmar to reclaim its place as a major rice bowl. The implementation of Thailand's much-criticized Paddy Pledging Scheme in 2012, which guarantees prices far above the market rate, has led Thai white rice to become uncompetitive internationally. As a result, Thailand slipped from its position as the world's top rice exporter in 2012 to trail behind India and Vietnam.[28]

Major food policy developments also took place at the regional level in 2012.[29] Regional cooperation on East Asian food security was enhanced during the Twelfth Meeting of the Association of Southeast Asian Nations (ASEAN) Ministers of Agriculture and Forestry and the Ministers of Agriculture of the People's Republic of China, Japan, and the Republic of Korea, which was held in Vientiane, Lao PDR, on September 28. The meeting reaffirmed the commitment to agricultural

Asia: International Goals Stimulate Small-Scale Farmers' Initiatives

MA. ESTRELLA A. PENUNIA

Four key policy developments in 2012 related to food systems and their governance affected small-scale women and men farmers in Asia. The first was the outcome document of the Rio+20 conference, titled "The Future We Want." While the document did not present many new ideas and its sustainable development goals lacked timeframes, we believed the document highlighted many goals that could stimulate and support the work of the Asian Farmers' Association for Sustainable Rural Development (AFA): promoting the right to food; making the role of small-scale farmers, fishers, pastoralists, and foresters significantly visible; ensuring food and nutrition security and upscaling sustainable agriculture systems; securing water access and sanitation; and using renewable sources of energy. With these reaffirmations from Rio+20 and some commitments from Asian governments, we can move on with our work, with more hope that our work will be supported by many stakeholders.

The second development was the adoption by the Committee on World Food Security's member states of the first version of the Global Strategic Framework for Food Security and Nutrition (GSF). The GSF is the result of a participatory, inclusive process involving civil society groups as active participants. It has many positive aspects, especially, again, its references to the right to food and to smallholder farmers, agriculture and food workers, small-scale fishers, pastoralists, indigenous people and the landless, women, and youth; and its recognition of agroecology. The challenge is for governments to provide funding to implement the GSF and for farmers' organizations to start their own initiatives and engage governments in GSF's implementation.

The third development was the adoption by Food and Agriculture Organization of the United Nations (FAO) member governments of Voluntary Guidelines on the Responsible Governance of Tenure of Land, Fisheries, and Forests in the Context of National Food Security. As small-scale farmers, fishers, pastoralists, and producers continue to struggle for their rights to natural resources and face more aggressive large-scale land acquisitions by both local elites and foreign companies and countries, this development gives us inspiration to press our governments to adopt these guidelines.

The last, equally important, development was the United Nations' declaration of 2012 as the International Year of Cooperatives—celebration of which helped highlight the importance of small-scale farmers' agricultural cooperatives in bringing about the farmers' economic empowerment—and of 2014 as the International Year of Family Farming.

In 2013, AFA's constituency will continue to face challenges with regard to securing our rights to land, water, forests, and seeds; making our governments enact favorable policies; ensuring that public-private partnerships in agricultural investments really put small-scale farmers at the center and empower them; and aligning public and private investments with sustainable but economically viable agro-technologies and enterprises that will also empower women and encourage young people to go into farming. With the hopes given by both the favorable policy documents in 2012 and the impending celebration of the Year of Family Farming in 2014, we will prepare ourselves for challenges and meaningful opportunities in 2013 by learning and sharing among ourselves and with others and partnering with allied groups and organizations who share the same interests and cause.

Ma. Estrella A. Penunia is secretary-general of the Asian Farmers' Association for Sustainable Rural Development, Quezon City, Philippines.

development within ASEAN+3 (ASEAN member states, the People's Republic of China, Japan, and the Republic of Korea), a commitment that first began in 1997.[30] This year also saw the ASEAN+3 Emergency Rice Reserve come into operation and release stockpiled rice in Indonesia, Lao PDR, and Thailand. The ASEAN+3 Emergency Rice Reserve is the product of a decade-long effort to

Thailand slipped from its position as the world's top rice exporter in 2012 to trail behind India and Vietnam.

improve upon the 33-year-old ASEAN Emergency Rice Reserve and involves both donated stocks and earmarked stocks.[31] However, research by the Asian Development Bank has suggested that, in the event of a calamity, the ASEAN+3 Emergency Rice Reserve would provide only temporary relief and that national governments would have to also turn to domestic reserves and new domestic policies.[32]

Research focusing on broad agricultural and agriculture-related policy issues will be critical to fostering both regional and national policies that contribute to agricultural and economic growth.

Kevin Z. Chen is a senior research fellow in the Development Strategy and Governance Division of the International Food Policy Research Institute, Beijing.

Latin America and the Caribbean

Eugenio Díaz-Bonilla

The recent economic performance of Latin America and the Caribbean (LAC) has been generally positive. Agricultural production and exports have grown over the past decade and generally remained strong in 2012, despite poor weather and a trade slowdown. Poverty has declined and the region's middle class has grown dramatically, as well. Nevertheless, economic inequality, particularly in land access, remains a major problem and source of conflict. The region also suffers from deforestation, although some steps have been taken to protect forests.

PRODUCTION

By the end of the 2000s, LAC had increased its agricultural production, and as a result its share of world agricultural production (valued in constant terms) had grown to almost 13 percent of global output (in the 1960s, LAC production accounted for less than 10 percent of global output). The region now accounts for a slightly larger share of world production than either the European Union or the United States and Canada combined. This increase resulted to a large extent from agricultural expansion in Brazil, but agricultural growth in other countries has contributed as well. Within that overall regional agricultural expansion, livestock production grew faster than crop production.

According to projections by the US Department of Agriculture (USDA), grain production is estimated to have dropped in Argentina and Mexico from about 85 to 75 million metric tons during 2011–2012. Oilseed production in Argentina and Brazil followed a similar pattern, falling from 133 to 116 million metric tons during the same period. Other countries, such as Bolivia, Ecuador, and Paraguay, also reported losses in grain production during 2012.[33] On the other hand, USDA projected that LAC would increase meat production in 2012: beef production was expected to increase by about 5 percent compared with 2011, and pork by about 6 percent (poultry production, by contrast, was expected to increase by less than 1 percent).

TRADE

During the last decade, LAC became the world's main net exporter of agricultural and food products, mainly, but not exclusively, because of the net trade surpluses generated by Argentina and Brazil. During 2012, there seems to have been a slowdown in world trade compared with 2011, but agricultural trade in general may have been less affected. Projections suggest that trade within the Americas continued to be the largest component of LAC trade flows, although for several agricultural products (such as soybeans), markets in Asia, and China in particular, have become the main destination. Several trade agreements began implementation in 2012, including those between the United States and Panama, the United States and Colombia, and a number of intraregional agreements. Also in 2012, Venezuela became member of the Southern Common Market (MERCOSUR).

PRODUCTIVITY AND RESEARCH AND DEVELOPMENT

Technological levels vary significantly between and within countries and across producer groups. Public investment in agricultural research and development (R&D) has increased somewhat in LAC, particularly over the last decade, but the region's average numbers are well below developed-nation levels. Moreover, a few countries account for most of the improvements, as investment has declined in the smaller and poorer countries that are most in need of agricultural R&D. During the last decades, the private sector—from multinational companies to producer associations—and civil society have also taken up active roles in the development and diffusion of agricultural technology.

In 2011, out of the 29 countries producing GM crops, 10 were in LAC. After the United States, the two countries with the largest areas devoted to GM crops that year were Brazil and Argentina. Also, Paraguay and Uruguay were among the top 10 countries that had more than 1 million hectares of GM crops in 2011. Although final numbers for the region as a whole are not available, Brazil appears to have increased its area under GM crops by 14 percent in 2012.[34]

LAND, DEFORESTATION, AND CLIMATE CHANGE

The gains in LAC's agriculture during the last decade, although driven in part by productivity improvements, also resulted from significant expansion of agricultural area: from 1990 to 2010, the region accounted for two-thirds of global deforestation. Unsurprisingly, land-use change contributes more to LAC's greenhouse gas emissions than any other source, though the region's emissions are comparatively low. Another worrisome consequence of rapid land-use change is the pressure this places on LAC's globally important reservoirs of biodiversity.

During the last years, however, several countries in LAC launched initiatives to reduce emissions resulting from deforestation and forest degradation. The annual rate of deforestation in the region, which is about three times higher than the world average, appears to have been reduced by about 20 percent in the last five years compared with the previous five-year period.[35]

An important development in this area during 2012 has been the approval of a controversial law aimed at reconciling production and conservation in Brazil. In May 2012, Brazilian President Dilma Rousseff enacted that law but rejected 12 articles (including an amnesty for illegal loggers) and made 32 modifications to the text. This partial veto was a compromise between farmers who argued that the original law was needed to increase food production and environmentalists who thought that it would lead to further destruction of the Amazon rainforest.

Hurricanes continued to affect Central America and the Caribbean in 2012. In particular, Hurricane Sandy in November caused several deaths and left thousands of people homeless in Haiti, while also exacerbating the cholera epidemic and causing extensive crop damage.

AGRARIAN STRUCTURE

LAC's agricultural performance has taken place against a background of large inequalities in land tenure, with small farms fragmenting and large landholdings expanding, all of which is squeezing out family farms and local communities. Land problems continued in 2012: in Honduras rural workers took part in coordinated land invasions across the country in April, and in Paraguay a land eviction in which 17 people were killed was used as an argument for the impeachment and subsequent removal of President Fernando Lugo in June. Conflicts around guerrilla activity and the cultivation of illicit drugs have also affected land patterns in several LAC countries. In this regard, a development with implications for opening land area to production was the start of exploratory talks between the Colombian government and guerrilla groups in the second half of 2012.

Honduras: Agricultural Extension and Better Technologies for Higher-Value Crops

JEREMÍAS VASQUEZ

The Association of Producers of Celeque (APROCEL) is a producers' association based in the Belen Gualcho municipality in the Department of Ocotepeque in Honduras. We were formed by 150 peasant families of Lenca origin that by culture and tradition produced only for subsistence: we grew beans and corn and cultivated small vegetable areas in the traditional way. In 2008, APROCEL, with the technical and financial support of the Global Village Project, the Jack & Marie Eiting Foundation, and the Millennium Challenge Account–Honduras, started a process of commercial production. This process involved appropriate technical assistance and the methodical implementation of a plan covering good agricultural practices, crop management, a nutrition program, and the adoption of the seed of more weather-resistant and high-yielding vegetable varieties. As our work progresses, we have continued to receive support from various national and international organizations.

In 2012, we saw a slight improvement in productivity in Honduras—not only in the plains, where most improvements traditionally take place, but also in the highlands—through the work of agricultural extension institutions that promote the use of seeds for adaptable varieties. This seed usage can lead to higher crop yields. Also in 2012, 75 families in APROCEL succeeded in adopting technologies—such as drip irrigation and protected cultivation in greenhouses or fiberweb macro tunnels—that improved the production infrastructure and had an extraordinary impact on productivity and family output. For this reason, we now focus on the production of high-value crops, such as vegetables—particularly lettuce, carrots, cabbage, tomatoes, potatoes, and chilies.

Currently we are implementing a step-by-step process within our organization for planned and competitive production for the formal and informal markets in northeastern Honduras. Our goal is to produce and sell about 1,500,000 pounds annually, with 70 percent of the production going to formal markets. Our vision for the medium and long term is to increase our productivity by incorporating and implementing agricultural best practices that allow us to gradually increase and stabilize our production. This process involves building on diversification and future value-addition to our products to help improve the incomes of the region's families.

In 2013, Central American countries and their governments should improve their policies and strategies—and not only regarding food security. They also should strengthen the productivity and competitiveness of the regions that have fewer opportunities and invest financial resources that are specifically targeted to stimulate the production systems for high-value crops. We know there is a lot of agroecological potential in our countries, and we should make better use of it to contribute in a meaningful and timely way to our countries' agrifood systems.

Jeremías Vasquez is a representative of the Association of Producers of Celeque, Belen Gualcho, Honduras.

OTHER ECONOMIC AND SOCIAL ISSUES

LAC economic growth slowed down in 2012, but positive trends continued overall. These trends, according to a recent study by the World Bank, have allowed the middle class to double in numbers during the last decade. In line with this development, poverty in LAC has declined in recent decades. It has also become more concentrated in urban centers, as rural populations and agricultural employment have fallen significantly. IFPRI's Global Hunger Index rates LAC more positively, on average, than any other developing region, except for Eastern Europe and the Commonwealth of Independent States. But of course, there are important differences across countries, and LAC continues to be one of the most unequal regions in the world according to different socioeconomic indicators. In particular, the region's land tenure structure is highly unequal, with Gini indexes in the 80s and 90s. Although 2012 data are not yet available, it is estimated that the trends mentioned have not been reversed.

The appreciation of real exchange rates in several LAC countries during 2011 and 2012 may put additional pressure on agricultural production's profit margins and agricultural trade in coming years by making it hard to export agricultural products and easier to import them.

During 2012, LAC's agriculture continued to support global food security (because LAC is the largest net exporter of agricultural and food products at the world level and helps stabilize food prices and supplies) and global environmental sustainability (because LAC is the largest provider, among developing regions, of global environmental goods, including biodiversity and oxygen). Also, the trends toward expansion of the middle class and reduction in poverty and food insecurity in the region seem to have continued. It remains to be seen what impact the continuation of global economic problems into 2013 and beyond will have on those trends. ■

Eugenio Díaz-Bonilla is a visiting senior research fellow at the International Food Policy Research Institute, Washington, DC.

Chapter 8

LOOKING AHEAD

Scenarios for the Future of Food

Mark W. Rosegrant, Simla Tokgoz, Prapti Bhandary, and Siwa Msangi

THE GLOBAL FOOD ECONOMY IS IN A STATE OF FLUX. A GROWING world population and rising incomes have pushed up demand for food, especially for more diverse diets. Food crops are being used to produce biofuels. Restrictive trade policies and speculation on food commodities have also contributed to the recent price spikes in global food markets.[1] Climate change, weather extremes, and natural resource degradation have created more challenging conditions for farmers. These factors are reflected in the higher and more volatile food prices of the past few years, after decades of stable and relatively low prices for major food commodities (Figure 1).[2]

Given the unsettled recent past, what lies ahead for the global food economy, especially in light of the rapid socioeconomic changes happening in emerging regions that will drive future diets and energy prices? Clearly, a whole range of policy choices and other factors will affect the world food system, and many of them will deal with sustainability issues raised at the 2012 United Nations Conference on Sustainable Development (Rio+20). These issues include sustainable agricultural production growth and the environmental impacts of human consumption patterns. The discussion of sustainable diets has been particularly strong within the European Union, where more consumers tend be highly conscious of their environmental footprint and where some forward-looking studies have closely examined the issue of meat consumption and its environmental impacts.[3] In this chapter, we address the implications of future consumption patterns beyond the European Union and explore the potential future impact of supply-side agricultural productivity improvements and higher energy prices on agriculture.

This chapter examines the dynamics of the new global food economy by looking at what would happen to global food prices, trade, and food security

Mark W. Rosegrant is the director of the Environment and Production Technology Division of the International Food Policy Research Institute (IFPRI) in Washington, DC. **Simla Tokgoz** is a research fellow in IFPRI's Markets, Trade, and Institutions Division. **Prapti Bhandary** is a senior research assistant in IFPRI's Environment and Production Technology Division. **Siwa Msangi** is a senior research fellow in IFPRI's Environment and Production Technology Division.

FIGURE 1 World prices for agricultural commodities, 1977–2012

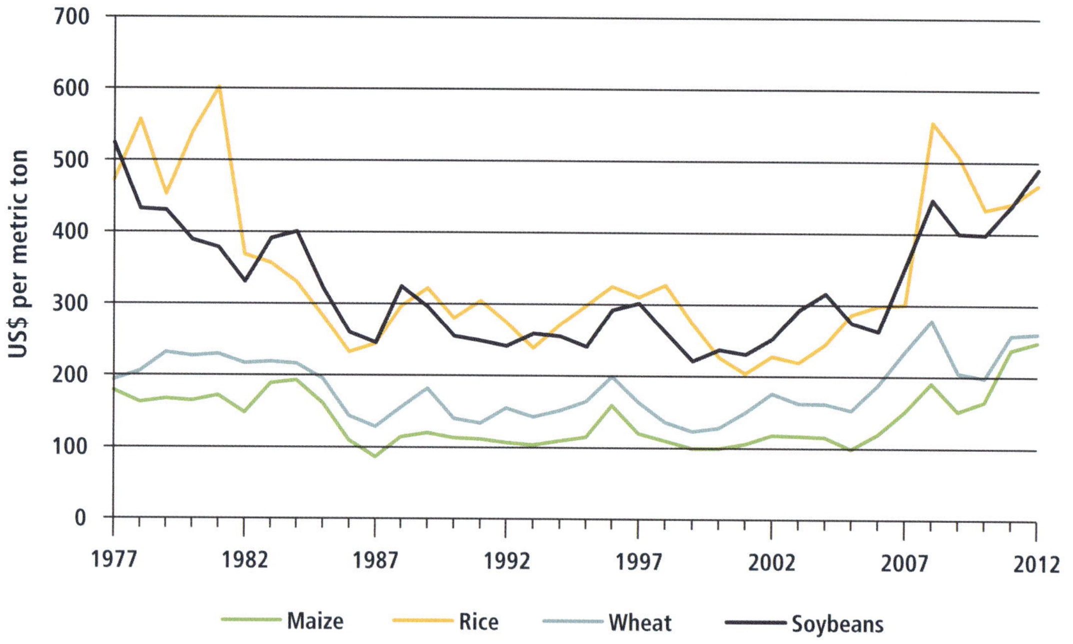

Source: World Bank, GEM Commodities, http://data.worldbank.org/data-catalog/commodity-price-data (accessed January 18, 2013). Prices for 2012 are through August 2012.
Note: Prices are in real 2005 US dollars.

under four alternative scenarios for the period between 2010 and 2050. To construct these scenarios, we used IFPRI's International Model for Policy Analysis of Agricultural Commodities and Trade (IMPACT), a model of the world food system that generates projections of global food supply, demand, trade, and prices.[4]

THE GROWING ROLE OF EMERGING ECONOMIES

In addition to showing global results, this chapter highlights results for Brazil, China, and India because, according to our simulations, growth in food demand and supply in these big emerging economies will be a major force behind future food security outcomes.

Brazil has become a huge player in global agricultural export markets. In two decades, the country has undergone an agricultural revolution to become a leading exporter of soybeans, beef, chicken meat, fruit, and pork, as well as biofuels. Brazil owes its agricultural advancement to investments in agriculture and related sectors, reduced trade restrictions, and macroeconomic policy reforms that generated economic stability. These policies triggered infrastructure development, advances in crop research, and adoption of innovative technologies, and gave the country a comparative advantage in the production of soybeans, sugar, meats, and other major crop commodities.[5]

China has experienced rapid agricultural growth but faces challenges as it seeks to meet the food needs of a growing population that is becoming wealthier and more urban. The country's future food security, trade position, and possible influence on international food markets are likely to depend on its future agricultural growth, its patterns of food demand growth, and the sensitivity of its trade balances and world prices to its environmental and growth outcomes.

In India, growth in agricultural productivity has been slow during the past two decades, and crop yields remain below those in most countries in Asia. At the same time, as their incomes rise, Indians are diversifying their diets. Demand is

Modeling the Future: How Can We Improve Food Policy?

GERALD NELSON

The outcomes of most decisionmaking processes, from how to allocate scarce research resources to when to implement new directions in national polices, have consequences that play out well into the future. Unexpected events can alter expected outcomes. The idea of strategic foresight activities (also called "scenario exercises") is to improve the payoff of decisionmaking by examining the range of potential outcomes. In the recent past, groups have used scenarios to explore many topics, including ecosystem challenges, energy futures, and water scarcity.

Strategic foresight activities use both qualitative and quantitative approaches to assess plausible futures and outcome ranges for decisionmaking. Quantitative models, such as the International Model for Policy Analysis of Agricultural Commodities and Trade (IMPACT) used in this chapter, have the advantage of explicitly specifying important relationships whose interactions are difficult to grasp qualitatively, clarifying the range of potential outcomes.

In 2012, the IFPRI-led CGIAR Research Program on Policies, Institutions and Markets began developing a strategic foresight platform,[1] which will build on work from the Global Futures for Agriculture project that evaluated promising technologies, investments, and policy reforms for simultaneously improving agricultural productivity and environmental sustainability.[2] The strategic foresight activities will augment the quantitative tools and methodologies from Global Futures to allow consistent assessments of potential research investments in agricultural technologies. The Policies, Institutions, and Markets program, involving all CGIAR centers and research programs, will institutionalize the following tools and techniques and extend them to national agricultural research centers and the private sector:

- Use existing software models that simulate crop productivity in varying environments to assess the effects of climate change at high spatial resolution.
- Simulate "virtual crops"—that is, characteristic-specific crops that have not yet been developed—to test their potential performance in different environments.
- Link results of models that simulate biological or physical changes of crop cultivation, water availability, and other factors with models that reflect the impact of such changes on the economy and society.
- Assess the potential socioeconomic consequences for human well-being if a new (virtual) crop variety were to become part of global agriculture.
- Allow users of a web-based application to customize scenarios to help answer their specific questions.

Despite these advances in technology, significant challenges remain. The existing crop models do not allow for assessments accounting for increased resistance to pests and diseases and do not incorporate the potential benefits of important changes in agricultural systems. The existing climate change datasets lack information on extreme weather events, and existing socioeconomic models do not sufficiently capture the complex interactions between agricultural and natural resource systems and the economy outside of agriculture. And, last but certainly not least, all of these analyses are based on woefully inadequate data. These are some of the challenges the CGIAR Research Program on Policies, Institutions and Markets must address to ultimately help policymakers decide which policy options can best establish future food security.

Gerald Nelson is a senior research fellow in the Environment and Production Technology Division of the International Food Policy Research Institute, Washington, DC.

growing for poultry and dairy products, in addition to traditional staples, fruits, and vegetables. India remains home to the highest number of food-insecure people.[6]

ALTERNATIVE SCENARIOS OF THE FUTURE

The analysis begins with a baseline scenario, which assumes that countries maintain their current trends in agricultural policies and investments from 2010 to 2050. We use this scenario as a basis for

FIGURE 2 Projected increase in world agricultural commodity prices, 2010–2050

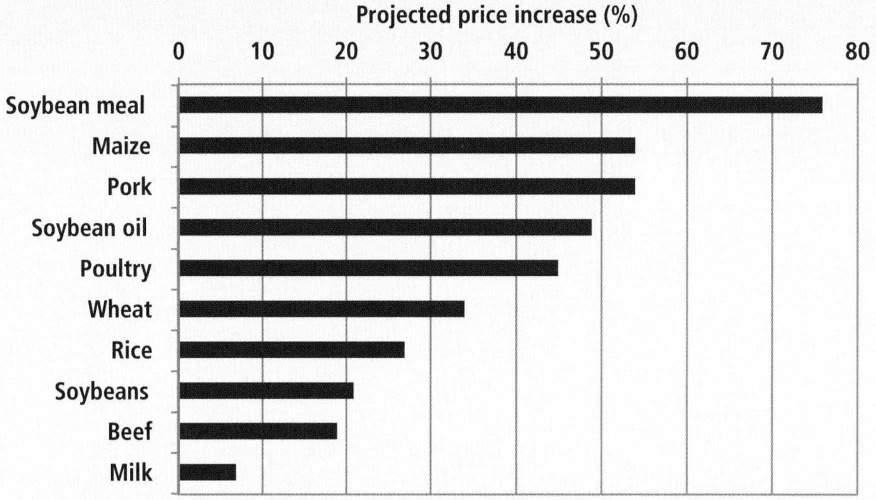

Source: IMPACT projections.

TABLE 1 Projected change in per capita food demand, 2010–2050 (%)

Region	Cereals[a]	Meat
East Asia and Pacific	-2	+48
Europe and Central Asia	+5	+17
Latin America and Caribbean	-4	+28
Middle East and North Africa	+4	+50
South Asia	+0	+143
Africa south of the Sahara	+24	+58
World	+3	+23

Source: IMPACT projections.
[a] : Projections do not include demand for cereals as livestock feed.

TABLE 2 Projected change in food security indicators, 2050 relative to 2010

Region	Number of malnourished children	Population at risk of hunger
East Asia and Pacific	-60	-33
Europe and Central Asia	-25	-13
Latin America and Caribbean	-50	-29
Middle East and North Africa	-50	+41
South Asia	-33	-31
Africa south of the Sahara	-7	+3
World	-29	-18

Source: IMPACT projections.

comparison with three other scenarios: (1) a higher-agricultural-productivity scenario, (2) a higher-energy-prices scenario, and (3) a lower-meat-demand scenario. Each of these scenarios represents a significant driver that can push or pull the trajectory of world food markets from either the demand or supply side. Although we could show what might happen through a conceptual or theoretical model, we have chosen a quantitative illustration to better convey the distribution of likely impacts and how quickly they might occur.

Projected rapid growth in meat and milk demand due to income and population growth and urbanization is projected to put pressure on prices for feedgrains and oilseed meals.

BASELINE SCENARIO: MAINTAINING CURRENT POLICIES

The major drivers of this baseline scenario are income growth, population increase, productivity gains in many agricultural activities, and biofuel sector expansion.[7] Projected rapid growth in meat and milk demand due to income and population growth and urbanization is projected to put pressure on prices for feedgrains and oilseed meals. Biofuel production will compete for land and water resources, according to the simulated results. Projected higher energy prices will increase the cost of production and make biofuels more competitive. Projected growing water and land scarcity will increasingly constrain growth in food production.

These dynamics, combined with the continuation of current policies and investments under this scenario, would result in higher projected world prices for agricultural commodities by 2050 (Figure 2). High demand for meat and livestock feed means that projected price increases would be greatest for soybean meal, maize, and pork. These rising prices reflect an assumption that most developing countries would be unable to rapidly meet growing domestic food demand. Thus major exporters would play a critical role in meeting the world's food consumption needs.

In most regions the baseline scenario projects a slightly increased demand for cereals as food (these results do not include demand for cereals as livestock feed). In Africa south of the Sahara, per capita demand for cereals would increase significantly (Table 1). In East Asia and the Pacific and Latin America and the Caribbean, however, income growth and changes in dietary patterns would lead to a projected decline in per capita demand for cereals. At the same time, income growth would cause projected meat demand to rise for all regions, and especially for East Asia and the Pacific (in absolute terms), and for South Asia (in percentage terms), where per capita income growth is projected to be highest.

As major exporters and importers, Brazil, China, and India would be critical forces in agricultural markets (Figure 5 on page 95). Under the baseline scenario Brazil would export a net 21 million metric tons of maize and 55 million metric tons of soybeans in 2050. At the same time, China would import a net 37 million metric tons of maize and 59 million metric tons of soybeans—clearly serving as an important market for Brazilian producers who are increasingly taking over from US producers.[8] China and India are projected to export significant quantities of rice.

Food security in IMPACT is measured through two indicators: the number of malnourished children and the population at risk of hunger.[9] Under the baseline scenario, food security would improve in most regions (Table 2). In Africa south of the Sahara, however, the decline in the number of malnourished children would be relatively slow. The reduction in the number of people at risk of hunger would also be slow for many regions, and in Africa south of the Sahara, as well as in the Middle East and North Africa, the number of people at risk of hunger would actually rise.

FIGURE 3 Projected difference in production compared with baseline scenario, 2050

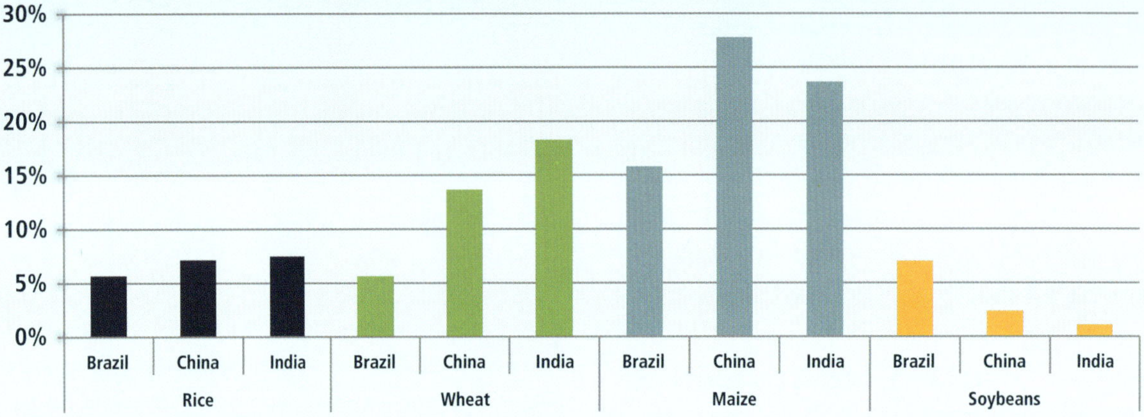

Source: IMPACT projections.

FIGURE 4 Projected difference in world commodity prices compared with baseline scenario, 2050

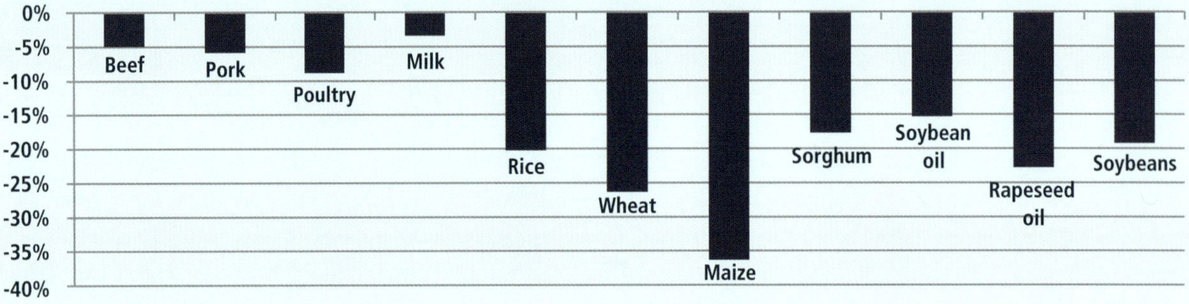

Source: IMPACT projections.

SCENARIO 1: HIGHER AGRICULTURAL PRODUCTIVITY

Recent increases in agricultural commodity prices have drawn attention to the importance of raising crop productivity and the role of agricultural research and development (R&D) in bringing about these productivity gains. Given that land use change is an important part of the discussion on the future sustainability of agricultural production, getting "more from less"—that is, expanding production by raising yields rather than by cultivating more land—will be a critical challenge. The higher-productivity scenario reflects agricultural investments that increase yields for major crops in all countries at rates that would keep inflation-adjusted crop prices in 2050 close to the level of the crop prices in 2010 in the baseline scenario. Figure 3 shows the changes in rice, wheat, maize, and soybean production in Brazil, China, and India that would be needed to keep prices at that level. Maize production would need to be 16–28 percent higher than in the baseline scenario, and wheat production would need to rise by 6–18 percent. These

FIGURE 5 Net trade in baseline and higher-productivity scenarios, 2050

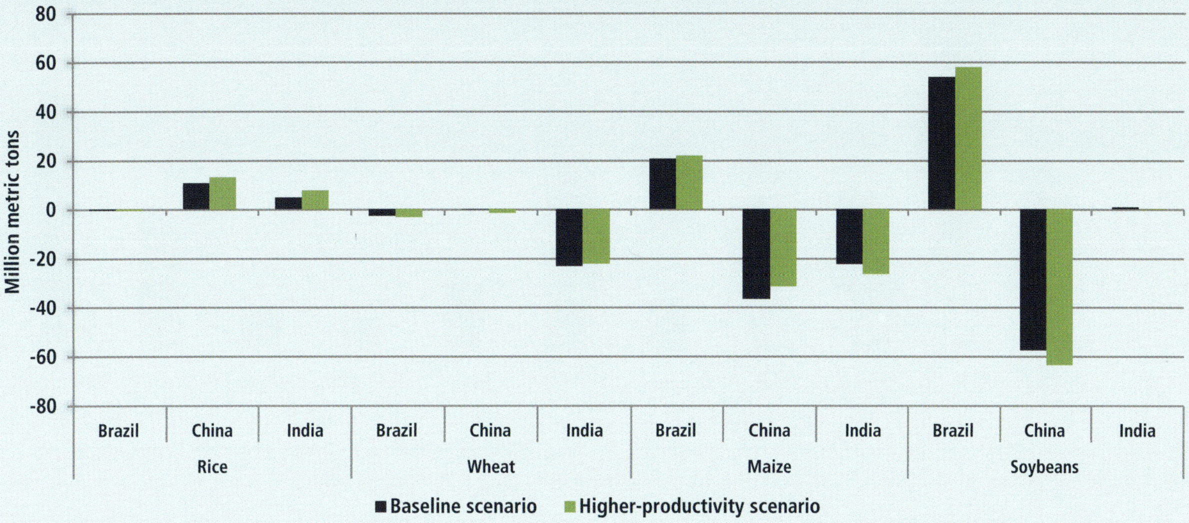

Source: IMPACT projections.

TABLE 3 Projected difference in food security indicators compared with baseline scenario, 2050 (%)

Region	Number of malnourished children	Population at risk of hunger
East Asia and Pacific	-9	-11
Europe and Central Asia	-13	-4
Latin America and Caribbean	-16	-19
Middle East and North Africa	-17	-16
South Asia	-5	-32
Africa south of the Sahara	-11	-32
World	-8	-24

Source: IMPACT projections.

findings underline the importance of investing in agriculture and agricultural R&D to keep food prices relatively low in the long term.

The scenario's projected higher yields would lower world commodity prices by increasing production and available supply in world markets. By 2050 cereal prices are projected to be 20–36 percent lower than in the baseline scenario (Figure 4). The simulated lower prices for cereals as feedstock would expand livestock and dairy production, causing meat and milk prices to fall. Lower bean and seed prices would lead to higher demand for oilseeds and greater production of vegetable oils, which would push down prices for soybean oil and rapeseed oil.

The higher-productivity scenario would also lead to changes in imports and exports of rice, wheat, maize, and soybeans by Brazil, China, and India, compared with the baseline scenario (Figure 5). Brazil plays a major role in soybean markets, along with the United States. In the higher-productivity scenario, it is projected that Brazil would increase its net exports of soybeans, and China would respond to Brazil's increased

FIGURE 6 Projected difference in world agricultural commodity prices compared with baseline scenario, 2050

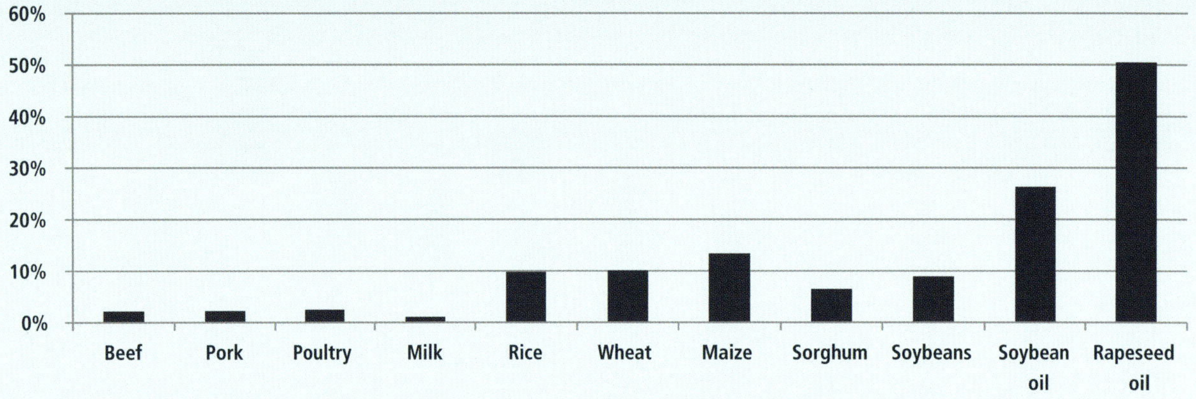

Source: IMPACT projections.

FIGURE 7 Net trade in baseline and higher-energy-prices scenarios, 2050

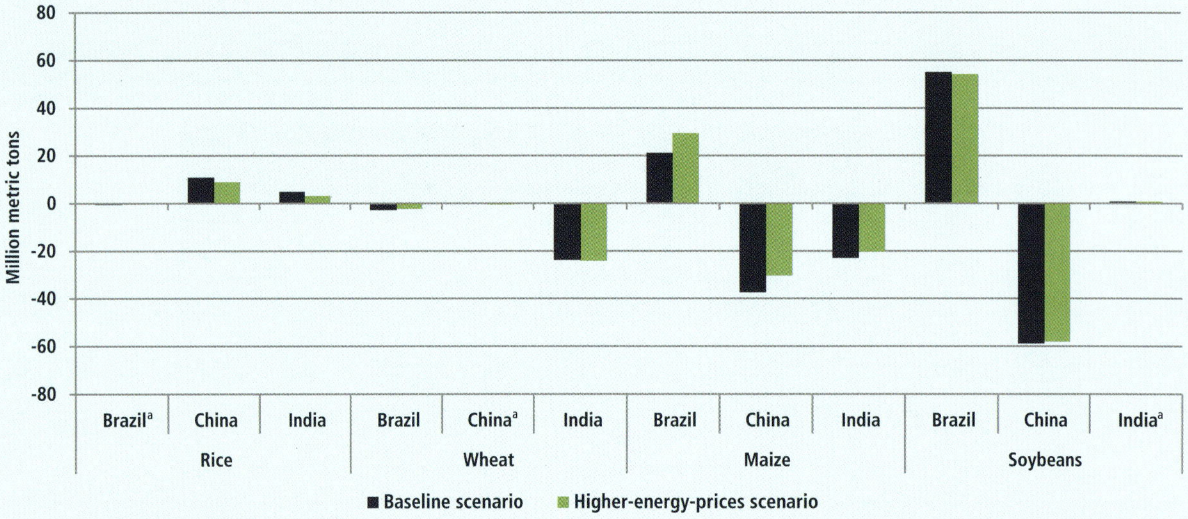

Source: IMPACT projections.
[a] Net trade is negligible.

TABLE 4 Projected difference in food security indicators compared with baseline scenario, 2050 (%)

Region	Number of malnourished children	Population at risk of hunger
East Asia and Pacific	+4	+6
Europe and Central Asia	+5	+2
Latin America and the Caribbean	+8	+17
Middle East and North Africa	+8	+8
South Asia	+2	+19
Africa south of Sahara	+4	+15
World	+4	+14

Source: IMPACT projections.

production by raising its net imports of soybeans. For rice, expanded Chinese and Indian exports would help increase supply in international markets and lower world prices, which would aid many developing countries that are net food importers. Brazil's net exports of maize would increase slightly. China, a major maize importer in 2050, would lower its imports with higher domestic supply, relieving some pressure on world markets. Changes in wheat trade would be relatively small. It is expected that Brazil and China would increase their net imports of wheat slightly, while India would decrease its net imports marginally.

The overall effect of the higher-productivity scenario would be to improve food security in all regions (Table 3). Higher yield growth that lowers agricultural commodity prices and raises food consumption leads to significantly lower numbers of malnourished children and people at risk of hunger. In fact, it is projected that the population at risk of hunger globally declines by 24 percent compared with the baseline scenario. Agricultural R&D and investment is clearly important not only for countries that can increase domestic production, but also for net importing countries, which benefit from productivity gains elsewhere.[10]

The overall effect of the higher-productivity scenario would be to improve food security in all regions.

SCENARIO 2: HIGHER ENERGY PRICES

One critical change in agricultural markets has been the recent strengthening of ties between the energy and agriculture sectors. The higher-energy-prices scenario aims to illustrate how energy prices affect agricultural production, even as agriculture provides energy products in the form of biofuel feedstocks. This scenario thus incorporates two links between agriculture and energy markets. The first link is energy prices' impact on biofuel demand and production. The scenario assumes a 100 percent increase in crude oil prices by 2035. Because higher oil prices make biofuel production more profitable, this assumption increases demand for feedstocks in the biofuel sector by an average of 67 percent for all countries and crops by 2035. (The scenario considers only first-generation biofuels made from, for example, maize, soybeans, and rapeseed, among other agricultural feedstocks.) The second link is energy prices' impact on fertilizer prices, which affect the cost of crop production. The scenario increases the annual growth rate of fertilizer prices by 75 percent throughout the period 2000–2050.

The large jump in energy prices in this scenario would significantly raise the prices of agricultural commodities, especially vegetable oils and maize (Figure 6), mainly because of the increased demand for crop-based feedstocks for biofuel production. Another factor is lower crop production due to the rise in production costs. Higher prices for livestock feed would push up the production costs of livestock and dairy producers and lead to higher meat and milk prices.

We also look in more detail at the scenario's impacts on imports and exports of rice, wheat, maize, and soybeans in Brazil, China, and India. Higher demand for soybean oil by Brazil's biodiesel industry would cause a projected drop in Brazilian soybean exports of 1 million metric tons compared with the baseline scenario (Figure 7). It is also projected that China would import 0.7 million metric tons fewer soybeans, and India would reduce its exports slightly. With the United States using its maize for its own domestic biofuel industry, Brazil would benefit from higher maize prices and expand its share of the maize market by increasing its net exports. China and India would lower their net imports of maize because higher world market prices would reduce demand. Rice would be less affected by the expansion of biofuel production. Nonetheless, by raising production costs, higher fertilizer prices would lead to reduced rice production and thus lower net exports from China and India. Brazil would lower its imports of wheat, whereas China and India would increase theirs.

The higher-energy-prices scenario would have serious consequences for food security. With higher prices and thus lower food consumption, the

FIGURE 8 Projected difference in world agricultural commodity prices compared with baseline scenario, 2050

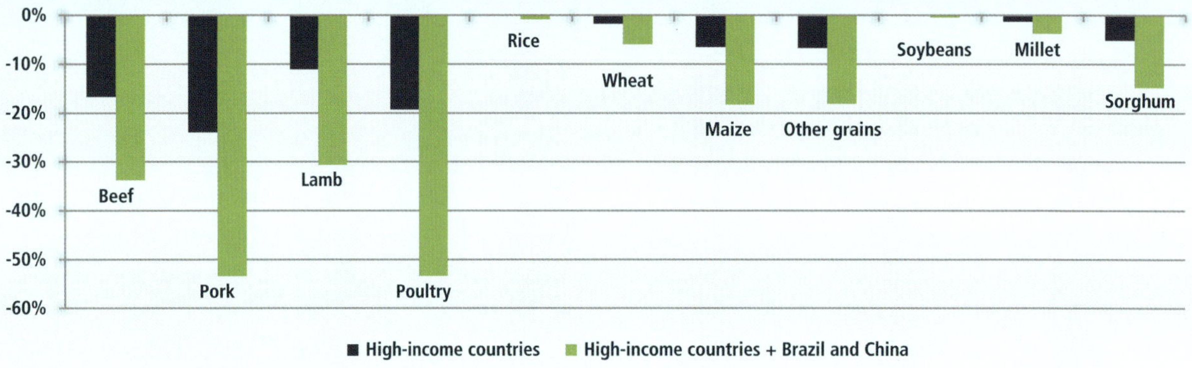

Source: IMPACT projections.

FIGURE 9 Projected difference in per capita meat demand compared with baseline scenario, 2050

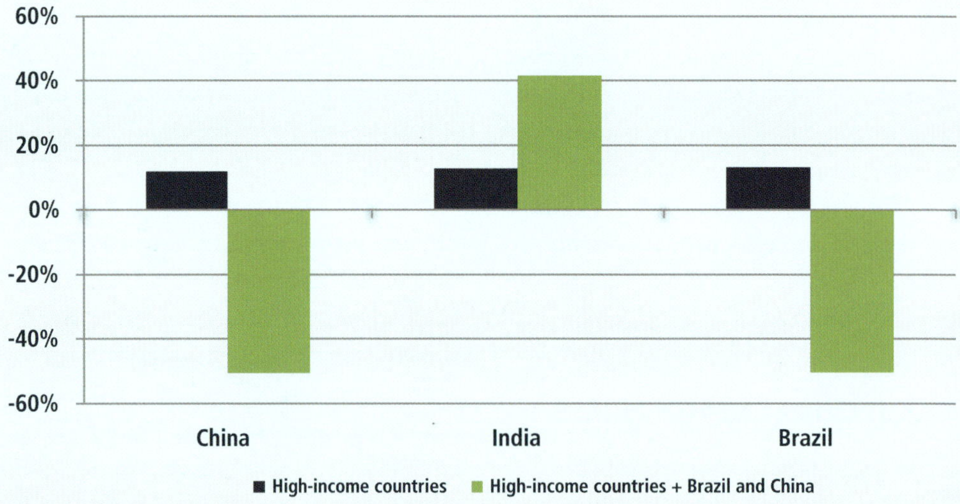

Source: IMPACT projections.

TABLE 5 Projected difference in food security indicators compared with baseline scenario, 2050 (%)

	Lower meat demand in high-income countries		Lower meat demand in high-income countries + Brazil and China	
Region	Number of malnourished children	Population at risk of hunger	Number of malnourished children	Population at risk of hunger
South Asia	-0.24	-1.6	-0.85	-5.6
Southeast Asia	-0.25	-0.8	-0.96	-2.8
Asia	-0.31	-1.1	-0.55	-3.7
Africa south of the Sahara	-0.96	-3.0	-2.94	-9.3
Latin America and Caribbean	-0.86	-0.9	-1.62	-1.9
World	-0.50	-1.8	-1.33	-5.5

Source: IMPACT projections.

number of malnourished children would increase in all regions in 2050 compared with the baseline (Table 4). The population at risk of hunger is also expected to increase for all regions and by 14 percent for the world as a whole.

SCENARIO 3: LOWER MEAT DEMAND

Dietary change is an important driver of demand for agricultural products that may have significant implications for the future sustainability of agriculture. The global structure of food demand is undergoing fundamental changes, driven largely by economic growth in developing countries. As consumers' incomes rise, they tend to shift their consumption from maize and other coarse grains to wheat and rice.[11] When urbanization further raises their incomes and leads to lifestyle changes, consumers make a secondary shift from rice to wheat—and their consumption of meat products also begins to increase.[12] The rise in consumption and production of meat leads to higher demand for coarse grains to be used for animal feed, rather than for direct human consumption.[13] In the developed, or high-income, countries, meat consumption is already very high, and growth in per capita meat and cereal consumption has slowed,[14] so these trends mean that developing countries will play a much larger role in global food markets.

Along with these fundamental shifts in long-term demand, strong concerns have been raised that meat-intensive diets in high-income countries put upward pressure on prices for cereals and coarse grains, as well as contributing to the high prevalence of chronic diseases[15] and increasing greenhouse gas emissions from the livestock sector.[16] Some analysts have asserted that cutting meat consumption in high-income countries, either through voluntary dietary changes or through policies such as taxes on livestock, would release cereals from livestock feed to food for poor people in developing countries.[17] The lower-meat-demand scenario examines whether reducing meat consumption in high-income countries and the emerging economies of Brazil and China would improve food security and reduce pressure on prices.

This scenario looks at two policy variations for cutting meat demand by 50 percent (1) in high-income countries and (2) in high-income countries plus Brazil and China. Brazil and China are included because they are important emerging economies, and they represent both a significant producer (Brazil) and a consumer (China) of meat products. These two countries play an important role in the world food demand and supply balance now and will likely do so in the future. (India is not considered here because it is not a major contributor to global meat demand.)

The global structure of food demand is undergoing fundamental changes, driven largely by economic growth in developing countries.

The simulation shows that if meat consumption were lowered in high-income countries, Brazil, and China, the drop in food prices would be notable (Figure 8). Given that this scenario focuses on reducing meat consumption, prices of livestock commodities would decline the most. But because demand for livestock feed would also decline, prices for cereals (particularly coarse grains such as maize) would drop as well.

What would happen to per capita meat demand in Brazil, China, and India under the lower-meat-demand scenarios? Our simulations show that lowering meat demand in high-income countries alone would make more meat available on world markets and lower prices, leading to higher meat consumption in Brazil, China, and India (Figure 9). If meat demand were reduced in high-income regions plus Brazil and China, Indian consumers would benefit from lower prices for animal products and livestock feed as well as from higher levels of meat consumption. (In India, of course, most meat consumption would be nonbeef.)

The lower-meat-demand scenarios would improve food security (Table 5). By reducing cereal prices and raising cereal consumption, these scenarios would lead to shifts in dietary preferences, increases in food availability, and eventually better nutritional status, particularly in developing countries.[18] If meat demand were cut in high-income countries alone, the number of malnourished children in Africa south of the Sahara would fall by about 1 percent compared with the baseline scenario. Reducing meat demand in Brazil and China as well would reduce the number of malnourished children in Africa south of the Sahara by nearly 3 percent and worldwide by about 1.3 percent.[19] Lower meat demand would also reduce the population at risk of hunger, with the largest drop occurring in Africa south of the Sahara.

CONCLUSIONS

Agricultural markets are undergoing a transformation. Long-run dynamics are changing demand-supply relationships, and new countries are emerging as major importers and exporters. Given these new realities, the four scenarios examined in this chapter show how different socioeconomic trends, policy actions, and investment choices could affect food prices and food security by 2050. In some cases, the results are dramatically different from the baseline scenario and call for closer attention on the part of market analysts, agricultural researchers, and policymakers.

> Different choices with regard to agricultural investment, energy, and food consumption can lead to vastly different results for food prices, trade, and food security.

One main takeaway message is that continuing upward pressure on food prices is likely to be the new norm, especially if trends in agricultural productivity growth continue to lag behind the demand-side drivers of change. If current agricultural, biofuel, and energy policies and trajectories of dietary change continue, inflation-adjusted prices of meat and grain are likely to rise. Although the baseline scenario projects fewer people at risk of hunger in most regions, it projects an increase in the Middle East and North Africa and Africa south of the Sahara.

Increased public and private investments in agricultural productivity growth, as in the higher-productivity scenario, would help to push agricultural commodity prices much lower than in the baseline scenario. Expanding funding for agricultural research and technology, extension services, rural infrastructure, irrigation, and water-use efficiency, among other efforts, can also lead to higher agricultural production and greater food security than in the baseline scenario.

Growing biofuel markets and the increasing share of energy in the costs of agricultural inputs such as fertilizer have intensified the linkage between energy and agricultural markets. Together with higher energy prices, this stronger linkage could make food prices even higher and more volatile than they have been in recent years, as shown in the higher-energy-prices scenario. The food-versus-fuel debate over biofuels therefore has critical repercussions for the food security of developing countries. Governments can undertake various policy initiatives to alleviate the pressures on food prices and food security, such as eliminating subsidies and trade barriers supporting crop-based biofuels.

Finally, in two lower-meat-demand scenarios, we consider whether reducing meat demand in developed countries is an effective route to improving food security in developing countries. A decline in consumption of livestock products in developed countries would have only small impacts on food security in developing countries. If reduced meat demand were extended to Brazil and China, the reduction in the number of malnourished children and the population at risk of hunger would be somewhat larger. More livestock products would be available in the global market, leading to lower prices in developing countries, where consumers could shift consumption to include more meat. Reduced demand for maize and other coarse grains for livestock feed would also tend to push down the prices of these commodities. However, this scenario would have little impact on the prices of wheat and rice—the main staple foods in most developing countries—and would therefore do little to raise consumption of these crops.

Although the scenarios included in this chapter are global in nature, special focus is given to Brazil, China, and India and their role in world agricultural markets. As significant producers and consumers, these three countries exert huge influence in agricultural markets. Changes in their dietary patterns, productivity growth, trade, and energy policies are likely to shape global trade patterns and therefore commodity prices. Although the United States and the European Union will remain important players in agricultural markets, the agricultural R&D efforts of Brazil, China, and India are becoming as important, if not more so, as those of the United States and the European Union.

These simulations of the global future show that different choices with regard to agricultural investment, energy, and food consumption can lead to vastly different results for food prices, trade, and food security. Higher investment in agricultural research that boosts productivity growth is projected to significantly improve the future food security situation. ■

FOOD POLICY INDICATORS: TRACKING CHANGE

TO DEVELOP AND IMPLEMENT EFFECTIVE FOOD POLICIES, DECISIONMAKERS AND POLicy analysts need solid evidence and timely information, among other things. As part of IFPRI's mission to provide research-based policy solutions that sustainably reduce poverty and end hunger and malnutrition, the Institute develops and shares global public goods, including datasets, indicators, and indexes. This information can be used to gauge the impact of policy changes and the progress on specific aspects of development. This section provides updates on data generated by IFPRI research in 2012, including indicators on investment in agricultural research, public spending on economic development in agriculture, global hunger levels, food policy research capacity, and total factor productivity.

Agricultural Science and Technology Indicators

Policymakers increasingly recognize greater investment in agricultural research as an essential element in raising agricultural productivity. Data on the size and scope of research capacity and investments, as well as on the changing institutional structure and functioning of agricultural research agencies, enhance our understanding of how agricultural research promotes agricultural growth. Indicators derived from such information allow the performance, inputs, and outcomes of agricultural research systems to be measured, monitored, and benchmarked.

IFPRI's Agricultural Science and Technology Indicators (ASTI) initiative is one of the few sources of statistics and other information on agricultural research in low- and middle-income countries. ASTI provides comprehensive quantitative and qualitative information on and analyses of investment, capacity, and institutional trends in agricultural research (see Table 1). ASTI data and analyses help research managers and policymakers improve policy formulation and decisionmaking at the national, regional, and international levels. All outputs are available on the ASTI website.

Website: www.asti.cgiar.org
Contact: Nienke Beintema
Email: asti@cgiar.org

TABLE 1 ASTI data

Low- and middle-income countries by region	Latest year available	Public agricultural research spending: 2005 PPP dollars (million)	Public agricultural research spending: 2005 US dollars (million)	Public agricultural research spending as a share of AgGDP (%)	Public agricultural researchers (FTEs)	Public agricultural researchers (FTEs) per million people economically engaged in agriculture	Share of women in public agricultural research staff (%)
Africa south of Sahara	2008	1,745	803	0.60	na	na	na
Benin	2008	21.6	9.0	0.57	115.4	73.7	16.9
Botswana	2008	18.8	8.9	4.32	97.4	318.1	29.8
Burkina Faso	2008	19.4	7.4	0.36	239.9	37.1	11.6
Burundi	2008	9.6	3.0	1.06	97.8	27.4	14.8
Congo, Republic of	2008	4.2	2.1	0.85	93.8	181.0	16.2
Côte d'Ivoire	2008	42.7	23.3	0.54	122.6	43.4	16.8
Eritrea	2008	3.0	1.2	0.74	121.9	82.9	32.0
Ethiopia	2008	68.7	17.9	0.26	1,318.3	43.7	6.8
Gabon	2008	1.6	0.8	0.20	61.4	324.9	22.4

Notes: na = not available. Table includes only countries where ASTI has conducted survey rounds since 2002. Data in italics are provisional estimates based on secondary data or macro-level survey rounds. Regional totals are from ASTI's latest global assessment (www.asti.cgiar.org/globaloverview). Public agricultural research includes government, higher-education, and nonprofit agencies, but excludes the private for-profit sector. Purchasing power parities (PPPs) measure the relative purchasing power of currencies across countries by eliminating national differences in pricing levels for a wide range of goods and services. PPPs are relatively stable over time, whereas exchange rates fluctuate considerably (for example, the fluctuations in the US dollar–euro rates of recent years). Measuring researchers in full-time equivalents (FTEs) takes into account the proportion of time researchers spend on research activities. For example, four university professors who spend 25 percent of their time on research would individually represent 0.25 FTEs and collectively be counted as 1 FTE. Regional totals and the overall total for low- and middle-income countries include estimates for non-ASTI countries. The totals therefore do not match the sum of the entries listed for specific countries. Where country data for 2008 were not available, they were extrapolated by various methods.
[a] Female shares are for 2006.

TABLE 1 Continued

Low- and middle-income countries by region	Latest year available	Public agricultural research spending: 2005 PPP dollars (million)	Public agricultural research spending: 2005 US dollars (million)	Public agricultural research spending as a share of AgGDP (%)	Public agricultural researchers (FTEs)	Public agricultural researchers (FTEs) per million people economically engaged in agriculture	Share of women in public agricultural research staff (%)
Gambia, The	2008	2.4	0.6	0.50	37.7	65.9	13.7
Ghana	2008	99.8	41.0	0.60	537.1	93.1	17.0
Guinea	2008	4.0	1.3	0.18	229.2	62.0	3.2
Kenya	2008	172.9	67.6	1.35	1,011.5	79.4	26.8
Madagascar	2008	12.0	3.9	0.29	212.4	31.3	29.8
Malawi	2008	20.8	6.9	0.71	126.5	27.3	15.7
Mali	2008	24.6	11.2	0.57	312.7	107.4	13.4
Mauritania	2008	6.3	2.3	0.63	73.7	104.1	4.9
Mauritius	2008	21.5	10.7	4.04	158.3	3,102.9	41.4
Mozambique	2008	17.7	8.4	0.37	263.3	31.6	29.0
Namibia	2008	21.5	14.5	1.96	70.2	267.9	16.5
Niger	2008	6.2	2.7	0.17	93.4	23.5	8.3
Nigeria	2008	403.9	185.3	0.43	2,062.0	168.1	21.3
Rwanda	2008	18.4	0.6	0.57	104.2	25.4	14.6
Senegal	2008	25.4	12.1	0.87	141.1	6.8	9.9
Seychelles	2008	0.2	0.2	0.70	1.0	39.1	na
Sierra Leone	2008	5.9	2.2	0.31	66.6	51.5	5.2
South Africa	2008	273.2	166.4	2.02	783.9	625.2	40.1
Sudan	2008	51.5	22.8	0.26	1,020.5	147.9	36.2
Tanzania	2008	77.3	27.1	0.58	673.5	42.0	21.3
Togo	2008	8.1	3.7	0.37	62.7	50.1	9.9
Uganda	2008	87.9	30.6	1.21	298.7	28.6	21.5
Zambia	2008	8.1	4.4	0.29	208.5	68.0	22.9
Zimbabwe	2008	na	na	na	147.7	47.6	43.2
Asia-Pacific	2008	7,725.1	3,319.7	0.42	na	na	na
Bangladesh	2009	125.9	44.3	0.31	2,081.3	64.6	15.8
Cambodia	2010	18.3	5.7	0.19	284.4	57.3	20.3
China	2008	4,047.6	1,703.0	0.50	43,200.0	85.7	na
India	2009	2,276.3	757.1	0.40	11,216.5	42.0	na
Indonesia	2009	*371.7*	*150.7*	*0.28*	na	na	na
Lao PDR	2003	10.4	2.9	0.24	123.4	56.9	22.3
Malaysia	2010	401.2	183.7	1.01	1,609.4	998.4	45.0
Myanmar	2003	4.6	4.6	0.06	618.7	32.6	54.2
Nepal	2009	22.3	7.1	0.23	388.7	33.2	9.5
Pakistan	2009	172.3	55.3	0.21	3,531.5	147.2	10.3

TABLE 1 Continued

Low- and middle-income countries by region	Latest year available	Public agricultural research spending: 2005 PPP dollars (million)	Public agricultural research spending: 2005 US dollars (million)	Public agricultural research spending as a share of AgGDP (%)	Public agricultural researchers (FTEs)	Public agricultural researchers (FTEs) per million people economically engaged in agriculture	Share of women in public agricultural research staff (%)
Papua New Guinea	2010	*21.0*	*9.1*	*0.43*	*163.1*	*77.3*	na
Philippines	2002	141.4	55.9	0.44	3,212.5	252.9	54.2
Sri Lanka	2009	37.5	13.1	0.34	618.8	154.3	34.8
Vietnam	2010	95.0	28.2	0.17	3,744.2	126.4	na
Latin America and Caribbean	2008	3,297.0	1,819.1	1.10	na	na	na
Argentina	2008	*441.1*	*192.8*	*0.93*	*3,930.5*	*2,781.7*	41.7[a]
Belize	2006	2.6	1.3	0.95	16.7	596.4	31.1
Brazil	2008	*1,402.6*	*781.7*	*1.52*	*4,633.2*	*398.7*	33.8[a]
Chile	2008	*103.1*	*61.4*	*1.38*	*674.6*	*696.2*	29.7[a]
Colombia	2008	*142.8*	*66.6*	*0.55*	*956.6*	*268.8*	31.7[a]
Costa Rica	2006	29.9	15.9	0.93	282.9	865.1	27.4
Dominican Republic	2006	17.4	10.3	0.26	138.8	278.7	24.9
El Salvador	2006	5.7	0.3	0.15	76.9	124.3	14.8
Guatemala	2006	8.3	4.2	0.06	102.4	53.2	14.7
Honduras	2006	12.7	4.7	0.43	123.7	181.3	7.4
Mexico	2008	*585.4*	*382.8*	*1.15*	4,066.8	*502.2*	22.3[a]
Nicaragua	2006	24.1	8.1	0.94	133.4	364.3	26.8
Panama	2006	10.0	5.7	0.50	166.7	653.6	16.1
Paraguay	2006	3.1	9.7	0.20	128.3	163.0	32.1
Uruguay	2006	59.8	32.4	1.99	400.4	2,107.6	42.5
Central and West Asia and North Africa	2008	1,848.4	824.5	0.68	na	na	na
Armenia	2010	1.8	0.7	0.07	294.9	1,966.0	44.3
Azerbaijan	2010	11.1	3.9	0.26	600.8	557.3	30.1
Iran, Islamic Republic of	2008	*731.3*	*218.2*	*0.96*	*4,890.9*	*753.7*	na
Jordan	2008	*6.8*	*3.6*	*1.59*	*212.7*	*1,849.1*	na
Kyrgyzstan	2010	1.7	0.5	0.09	160.5	313.4	26.5
Morocco	2008	*96.2*	*52.9*	*0.58*	*463.0*	*150.4*	na
Syrian Arab Republic	2008	*74.1*	*130.2*	*0.45*	*1,888.1*	*1,501.5*	na
Tunisia	2002	51.2	22.9	0.91	440.8	574.0	27.6
TOTAL LOW- AND MIDDLE-INCOME	2008	15,578.3	7,238.3	0.54	na	na	na

Statistics of Public Expenditure for Economic Development

The Statistics of Public Expenditure for Economic Development (SPEED) database is an IFPRI resource that contains information on agricultural and other sectoral public expenditures in 112 developing countries and 34 developed countries from 1980 to 2010 (see Table 2). IFPRI researchers have compiled data from multiple sources, including the International Monetary Fund, the World Bank, and national governments, and conducted extensive data checks and adjustments to ensure consistent spending measurements over time that are free of exchange-rate fluctuations and currency denomination changes. Differences from the data in the *2011 Global Food Policy Report* may arise from the revision of public expenditure data and other variables such as population and the gross domestic product deflator. The SPEED database is updated periodically.

Policymakers, researchers, and other stakeholders can use this robust database for many purposes. The data allow users to examine historical trends and the allocation of government resources across sectors and make comparisons with other countries within a region or at a similar level of development. They also aid in analyzing the alignment of actual expenditure with broad development priorities for economic growth, poverty reduction, and food security. Such analysis helps users to monitor progress toward achieving development goals, identify funding gaps, align development strategy priorities with national budget execution, and diagnose the efficiency and cost-effectiveness of public spending.

Website: www.ifpri.org/ourwork/programs/priorities-public-investment/speed-database

Contact: Bingxin Yu (b.yu@cgiar.org)

TABLE 2 Agricultural public expenditure for economic development, by country

	Agricultural expenditure (billions 2005 constant US dollars)				Agricultural expenditure (billions 2005 PPP dollars)				Per capita agricultural expenditure (2005 constant US dollars)				Per capita agricultural expenditure (2005 PPP dollars)				Ratio of agricultural expenditure to agricultural GDP (%)				Share of agriculture in total expenditure (%)			
Region/country	1980	1995	2010	*	1980	1995	2010	*	1980	1995	2010	*	1980	1995	2010	*	1980	1995	2010	*	1980	1995	2010	*
East Asia and Pacific																								
China	7.1	8.9	88.9		17.0	21.1	211.3		7.2	7.3	65.3		17.2	17.3	155.2		10.9	4.7	24.3		12.2	8.4	10.3	
Fiji	0.03	0.03	0.02	2	0.04	0.03	0.02	2	54.7	36.2	20.9	2	64.7	42.8	24.8	2	8.5	5.8	4.0	2	7.2	4.0	2.0	2
Indonesia	1.8	1.8	1.6	3	4.4	4.4	3.9	3	12.3	9.4	7.0	3	30.3	23.1	17.2	3	9.3	4.8	3.5	3	10.3	5.9	2.6	3
Malaysia	0.8	1.0	2.9		1.6	2.1	6.3		54.8	47.2	103.7		119.8	103.0	226.6		10.8	8.7	17.7		8.8	5.1	6.7	
Mongolia		0.01	0.02			0.0	0.1			4.5	8.8			13.1	25.5			1.9	3.8			2.8	2.4	
Myanmar	19.2	12.4	18.3	3	0.4	0.3	0.4	3	571.8	282.4	372.4	3	13.2	6.5	8.6	3	8.0	2.7	1.4	3	23.6	14.9	8.3	3
Papua New Guinea	0.1	0.1	0.0	3	0.2	0.1	0.1	3	24.3	10.7	3.7	3	56.4	24.8	8.7	3	8.2	3.1	1.2	3	8.5	4.0	1.5	3
Philippines	0.4	0.9	1.3		1.1	2.2	3.3		8.8	12.2	13.9		22.4	31.0	35.1		3.2	5.7	8.0		6.1	6.9	5.9	
Singapore	0.02	0.03	0.06		0.0	0.0	0.1		8.4	8.0	13.1		12.9	12.4	20.3		5.4	23.9	102.3		0.4	0.2	0.2	
Thailand	0.8	2.4	2.5		1.9	6.1	6.4		16.1	39.9	37.2		40.6	100.7	94.0		7.8	18.8	9.7		9.7	11.3	5.8	
Tonga	0.005	0.001	0.004		0.008	0.001	0.007		53.9	5.3	43.3		87.0	8.6	69.8		9.5	1.1	8.6		10.0	0.8	3.5	
Vanuatu	0.001	0.003	0.003	3	0.003	0.005	0.006	3	12.2	16.4	13.1	3	22.8	30.7	24.6	3	4.4	5.4	3.2	3	3.0	3.1	5.0	3
Vietnam		0.4	1.0			1.5	3.3			6.0	11.1			20.3	37.5			6.1	6.7			8.2	3.9	
South Asia																								
Afghanistan			0.1				0.3				2.8				9.0				2.4				3.7	

Note: PPP (purchasing power parity) dollars measure the relative purchasing power of currencies across countries by eliminating national differences in pricing levels for a wide range of goods and services. Because of the dramatic differences in countries' agriculture spending, entries have different numbers of decimal places.
* 1 = last year of data available is 2009; 2 = last year of data available is 2008; 3 = last year of data available is 2007; 4 = last year of data available is 2006.

TABLE 2 Continued

	Agricultural expenditure (billions 2005 constant US dollars)				Agricultural expenditure (billions 2005 PPP dollars)				Per capita agricultural expenditure (2005 constant US dollars)				Per capita agricultural expenditure (2005 PPP dollars)				Ratio of agricultural expenditure to agricultural GDP (%)				Share of agriculture in total expenditure (%)			
Region/country	1980	1995	2010	*	1980	1995	2010	*	1980	1995	2010	*	1980	1995	2010	*	1980	1995	2010	*	1980	1995	2010	*
Bangladesh	0.2	0.2	0.9	1	0.5	0.7	2.4	1	2.0	1.9	5.2	1	5.6	5.3	14.9	1	2.9	2.6	6.2	1	13.0	4.9	8.9	1
Bhutan	0.01	0.03	0.04	1	0.03	0.09	0.12	1	27.8	61.8	61.4	1	77.8	173.2	172.0	1	19.3	22.6	21.1	1	31.9	19.7	11.2	1
India	1.8	3.4	13.0		5.3	10.2	39.2		2.6	3.6	10.7		7.7	10.7	32.2		2.4	2.9	6.4		7.2	5.3	6.8	
Maldives	0.00	0.02	0.01		0.01	0.04	0.01		25.5	90.9	26.1		40.2	143.0	41.0		10.8	38.5	17.6		8.8	12.1	1.3	
Nepal	0.1	0.1	0.2		0.2	0.3	0.5		4.2	4.1	5.8		13.3	12.8	18.2		3.8	3.8	5.1		16.4	9.6	8.6	
Pakistan	0.1	0.1	0.8		0.4	0.2	2.5		1.4	0.6	4.4		4.5	1.8	13.6		1.3	0.4	2.8		2.1	0.5	3.1	
Sri Lanka	0.2	0.2	0.40	2	0.5	0.7	1.14	2	12.7	13.4	19.9	2	36.3	38.2	56.9	2	8.7	6.7	10.0	2	5.8	5.3	6.1	2
Europe and Central Asia																								
Albania		0.1				0.1				18.9				43.3				2.2				3.9		
Azerbaijan		0.1	0.2			0.2	0.7			10.9	27.0			31.0	76.9			6.2	11.0			8.0	3.1	
Belarus		0.1	1.0			0.4	2.9			12.6	107.6			34.8	297.4			4.7	26.8			4.2	11.0	
Bulgaria		0.02	0.14			0.0	0.4			1.9	18.5			5.1	48.8			0.5	7.8			0.3	1.6	
Georgia			0.01				0.03				2.6				6.3				1.3				0.4	
Kazakhstan			0.7				1.5				41.8				96.4				15.8				4.2	
Kyrgyzstan		0.02	0.02	3		0.1	0.1	3		3.3	3.0	3		12.0	10.8	3		2.2	1.9	3		3.5	2.4	3
Latvia		0.1	0.3			0.2	0.5			52.3	130.7			97.9	244.6			17.5	45.6			5.4	5.2	
Lithuania		0.3	0.3			0.5	0.5			77.3	85.9			144.4	160.5			17.9	29.2			8.7	2.7	
Moldova		0.01	0.04			0.03	0.12			2.7	11.4			7.7	32.5			1.5	12.1			1.4	3.2	
Romania	2.7	2.0	1.6	3	5.6	4.0	3.2	3	122.4	86.3	73.5	3	250.9	176.8	150.6	3	23.1	11.4	15.9	3	7.4	7.7	4.9	3
Russian Federation		0.2	0.8			0.5	1.7			1.5	5.3			3.2	11.8			0.6	1.8			0.2	0.3	
Serbia			0.2				0.6				23.0				57.9				6.0				3.0	
Ukraine			0.6				1.8				13.3				40.5				8.2				2.4	
Middle East and North Africa																								
Algeria		0.5	1.7	1		1.2	4.0	1		18.6	49.2	1		42.8	113.4	1		7.2	13.0	1		2.4	3.6	1
Bahrain	0.01	0.01	0.01	2	0.0	0.0	0.02	2	24.9	20.4	16.4	2	37.7	30.9	24.8	2	17.5	16.9	23.3	2	0.6	0.6	0.3	2
Egypt	0.7	0.9	0.9	3	2.4	3.2	3.1	3	14.9	14.1	10.9	3	53.2	50.4	39.0	3	12.0	8.8	5.8	3	5.1	5.0	3.0	3
Iran, Islamic Republic of	1.0	1.1	0.8	1	3.3	3.6	2.6	1	24.8	17.2	10.5	1	83.1	57.7	35.3	1	7.2	4.9	3.2	1	3.4	4.2	1.4	1
Jordan	0.02	0.11	0.04		0.0	0.2	0.1		8.3	25.5	6.6		15.5	47.6	12.3		5.1	32.2	8.7		1.0	4.5	0.8	
Kuwait	0.01	0.12	0.15	1	0.0	0.2	0.2	1	7.6	70.8	50.5	1	10.3	96.8	69.0	1	14.3	58.8	53.5	1	0.1	0.5	0.3	1
Lebanon		0.02	0.02	1		0.04	0.04	1		6.5	4.8	1		11.5	8.6	1		1.8	1.4	1		0.4	0.2	1
Morocco	0.5	0.5	0.4	3	1.0	1.0	0.7	3	27.7	19.9	12.8	3	50.3	36.1	23.2	3	11.6	9.2	4.4	3	6.8	4.5	2.0	3
Occupied Palestinian Territory			0.02	1			0.03	1			3.7	1			8.1	1			4.5	1			0.6	1
Oman	0.0	0.1	0.1	3	0.1	0.2	0.1	3	42.0	58.6	23.2	3	69.5	97.0	38.4	3	27.9	20.8	13.8	3	1.8	1.6	0.5	3
Syrian Arab Republic	0.3	0.5	0.4	1	0.7	1.4	1.2	1	30.1	35.3	19.8	1	79.7	93.4	52.5	1	11.3	7.9	5.8	1	5.0	10.2	4.7	1
Tunisia	0.5	0.4	0.5		1.2	1.0	1.1		82.1	49.9	49.1		183.3	111.5	109.5		28.1	17.2	15.8		15.6	8.3	5.5	
Turkey	0.6	0.6	6.78	2	1.0	0.9	10.96	2	13.5	9.4	91.7	2	21.8	15.2	148.3	2	1.4	1.1	14.5	2	2.1	0.8	5.1	2
United Arab Emirates	0.06	0.05	0.05	1	0.1	0.1	0.1	1	56.3	21.7	11.6	1	84.7	32.7	17.5	1	14.0	2.8	2.7	1	0.8	0.7		3
Yemen	0.02	0.04	0.07	3	0.0	0.1	0.2	3	2.1	2.5	3.3	3	5.7	6.8	9.0	3		1.8	3.9	3	1.4	1.7	1.1	3

TABLE 2 Continued

Region/country	Agricultural expenditure (billions 2005 constant US dollars) 1980	1995	2010	*	Agricultural expenditure (billions 2005 PPP dollars) 1980	1995	2010	*	Per capita agricultural expenditure (2005 constant US dollars) 1980	1995	2010	*	Per capita agricultural expenditure (2005 PPP dollars) 1980	1995	2010	*	Ratio of agricultural expenditure to agricultural GDP (%) 1980	1995	2010	*	Share of agriculture in total expenditure (%) 1980	1995	2010	*
Latin America and Caribbean																								
Argentina	0.2	0.1	0.9	1	0.4	0.3	2.1	1	6.0	4.1	22.3	1	13.8	9.5	51.1	1	2.2	1.7	5.2	1	0.6	0.6	1.7	1
Bahamas	0.01	0.02	0.01		0.01	0.02	0.02		54.5	65.2	41.9		70.4	84.3	54.2		12.1	9.6	11.9		1.5	1.7	1.0	
Barbados	0.02	0.03			0.04	0.04			90.7	106.4			146.6	172.0			9.9	17.1			3.2	2.8		
Belize	0.01	0.01			0.02	0.01			64.4	33.1			105.5	54.2			11.3	6.4				4.6		
Bolivia (Plurinational State of)	0.03	0.00	0.03	3	0.11	0.01	0.11	3	5.6	0.4	3.2	3	20.4	1.6	11.5	3	2.8	0.3	2.2	3	3.3	0.3	1.4	3
Brazil		7.2	5.5			13.0	9.8			44.7	27.9			80.2	50.0			18.0	8.3			5.7	2.2	
Chile	0.2	0.2	0.4		0.3	0.3	0.7		16.7	12.3	24.4		28.1	20.6	40.9		7.2	2.4	9.0		1.8	1.2	1.3	
Colombia	0.2	0.3	0.5		0.4	0.6	1.2		6.6	7.5	11.6		14.2	16.1	24.9		1.3	1.5	4.3		3.7	1.8	0.5	
Costa Rica	0.1	0.0	0.2		0.1	0.1	0.3		27.6	11.3	34.5		53.8	22.0	67.4		4.7	2.2	9.3		4.7	2.3	3.3	
Dominican Republic	0.3	0.2	0.1	3	0.5	0.3	0.2	3	50.3	23.0	13.4	3	88.9	40.6	23.7	3	12.0	9.3	5.0	3	16.7	9.0	1.9	3
Ecuador	0.2		0.2		0.5		0.5		24.5		15.1		58.0		35.8		8.2		6.5		7.3		1.6	
El Salvador	0.10	0.03	0.06		0.20	0.06	0.12		21.5	5.3	9.4		43.4	10.7	18.9		4.0	1.6	2.5		5.8	1.7	1.8	
Grenada		0.01				0.02				118.0				174.4				33.7				9.7		
Guatemala	0.14	0.05	0.10		0.26	0.09	0.18		19.5	4.7	6.8		37.0	8.9	12.9		3.8	1.0	2.5		7.9	2.7	2.0	
Jamaica		0.1	0.1	1		0.1	0.2	1		31.7	36.2	1		53.0	60.4	1		8.4	13.8	1		2.3	2.3	1
Mexico	8.4	3.2	4.7		12.9	4.9	7.2		122.7	34.6	42.6		187.6	53.0	65.2		20.4	9.4	12.4		14.6	4.1	2.3	
Panama	0.11	0.04	0.08	1	0.21	0.08	0.15	1	55.2	15.1	22.8	1	106.0	29.0	43.9	1	18.1	5.1	6.2	1	5.3	1.6	1.4	1
Paraguay	0.01				0.04				4.5				13.8				1.2				3.5			
Peru			0.2				0.5				7.5				16.7				2.9				1.6	
Saint Vincent and the Grenadines	0.002	0.004	0.003	3	0.004	0.006	0.005	3	20.9	33.4	28.3	3	36.5	58.4	49.5	3	7.7	7.1	7.7	3	3.8	3.5	2.1	3
Trinidad and Tobago	0.2	0.1	0.11	2	0.3	0.2	0.17	2	143.0	85.2	79.1	2	236.1	140.6	130.6	2	69.6	52.3	153.6	2	5.1	4.5	1.7	2
Uruguay	0.0	0.0	0.1	3	0.1	0.1	0.1	3	12.4	13.9	18.7	3	22.9	25.7	34.4	3	2.0	3.5	3.1	3	2.3	2.4	1.5	3
Venezuela (Bolivarian Republic of)			0.5	3			0.9	3			18.3	3			33.1	3			7.3	3			1.1	3
Africa south of Sahara																								
Angola		0.1	0.41	2		0.1	0.80	2		5.4	22.6	2		10.5	44.2	2		6.7	12.1	2		1.7	2.3	2
Benin			0.05	2			0.12	2			5.6	2			13.5	2			3.1	2			4.6	2
Botswana	0.0	0.1	0.2	1	0.1	0.3	0.3	1	50.7	76.8	81.5	1	107.1	162.1	172.1	1	21.0	49.1	45.7	1	9.7	6.0	3.0	1
Burkina Faso	0.1	0.3	0.2		0.3	0.7	0.5		18.5	26.9	12.2		48.9	70.8	32.1		25.4	26.3	8.5		31.3	45.2	10.8	
Burundi		0.01				0.04				2.0				6.2				3.3				4.6		
Cameroon	0.03	0.05	0.11	4	0.06	0.11	0.24	4	3.3	3.8	6.1	4	7.0	7.9	12.9	4	1.1	2.0	3.3		2.2	4.2		
Cape Verde		0.002	0.017			0.00	0.02			4.3	32.9			5.5	42.1			2.2	13.6				2.8	1
Central African Republic	0.03	0.02	0.01		0.06	0.04	0.01		12.6	5.8	1.5		25.3	11.5	3.1		6.6	3.3	0.8				2.3	
Congo		0.00				0.01				1.8				3.5				1.1				0.3		
Côte d'Ivoire	0.13	0.12	0.08	1	0.24	0.22	0.14	1	15.5	7.9	3.7	1	28.5	14.6	6.8	1	4.2	3.4	1.8	1	2.6	3.5	2.1	1
Democratic Republic of the Congo		0.11	0.04	3		0.24	0.10	3		2.4	0.7	3		5.3	1.6	3		2.7	1.3	3		0.2	1.8	3
Equatorial Guinea		0.01	0.05	1		0.01	0.09	1		14.9	68.6	1		27.4	125.9	1		2.3	12.4	1			0.8	1

TABLE 2 Continued

Region/country	Agricultural expenditure (billions 2005 constant US dollars) 1980	1995	2010	*	Agricultural expenditure (billions 2005 PPP dollars) 1980	1995	2010	*	Per capita agricultural expenditure (2005 constant US dollars) 1980	1995	2010	*	Per capita agricultural expenditure (2005 PPP dollars) 1980	1995	2010	*	Ratio of agricultural expenditure to agricultural GDP (%) 1980	1995	2010	*	Share of agriculture in total expenditure (%) 1980	1995	2010	*
Ethiopia	0.06	0.12	0.56	1	0.22	0.46	2.16	1	1.7	2.1	6.9	1	6.3	8.0	26.6	1	1.7	2.9	6.0	1	7.0	9.7	17.5	1
Gambia	0.01				0.06				24.2				91.5				24.0				17.1			
Ghana	0.06	0.02	0.01	3	0.14	0.04	0.03	3	5.4	1.0	0.6	3	13.1	2.5	1.4	3	2.2	0.6	0.4	3	12.2	0.7	0.5	3
Guinea-Bissau		0.00				0.00				0.1				0.2				0.0						
Kenya	0.18	0.27	0.29		0.47	0.70	0.75		11.2	10.0	7.2		28.8	25.6	18.3		6.4	6.3	6.4		8.3	7.0	4.6	
Lesotho	0.01	0.07	0.03	2	0.03	0.13	0.06	2	11.5	39.9	15.0	2	21.0	72.8	27.4	2	10.2	37.2	25.5	2	8.0	12.4	3.2	2
Liberia	0.029	0.004	0.005	2	0.059	0.008	0.011	2	15.3	1.9	1.4	2	31.1	3.9	2.8	2	4.9	3.1	1.3	2	5.0	2.8	3.4	2
Madagascar		0.04	0.09	2		0.12	0.26	2		3.0	4.5	2		9.3	13.8	2		4.0	5.7	2		6.1		2
Malawi	0.05	0.05	0.25	1	0.15	0.16	0.76	1	8.1	5.2	16.5	1	24.3	15.7	49.6	1	8.0	8.5	22.5	1	10.2	8.8	23.2	1
Mali	0.000	0.001	0.002	2	0.001	0.003	0.004	2	0.0	0.1	0.1	2	0.1	0.3	0.3	2	0.0	0.1	0.1	2	8.3	16.3	12.7	2
Mauritius	0.03	0.05	0.08		0.07	0.10	0.15		34.6	44.0	58.8		69.6	88.5	118.1		14.2	11.7	23.0		6.9	5.5	3.8	
Mozambique			0.11				0.23				4.6				9.8				3.6				3.6	
Namibia		0.10	0.11	3		0.15	0.16	3		63.4	50.6	3		94.6	75.4	3		17.7	13.8	3		6.0	4.7	3
Niger	0.05	0.05	0.12	1	0.13	0.13	0.29	1	9.1	5.8	8.1	1	21.3	13.5	18.9	1	5.3	5.5	8.4	1	13.2	13.9	12.1	1
Nigeria	0.51	0.22	0.89		1.12	0.49	1.95		6.9	2.0	5.6		15.0	4.4	12.3		3.2	1.0	1.5		2.9	3.6	4.4	
Rwanda			0.07				0.22				7.1				21.2				5.8				6.6	
Senegal	0.04	0.05	0.39		0.09	0.11	0.82		7.2	6.1	30.3		15.2	12.8	63.5		5.1	4.4	22.7		4.3	5.2	13.9	
Seychelles		0.01	0.01			0.01	0.01			96.3	68.6			156.8	111.6			26.1	27.0			2.0	1.4	
Sierra Leone		0.00	0.01	3		0.01	0.01	3		0.5	1.0	3		1.4	2.6	3		0.6	0.7	3		1.6		
South Africa		0.25	0.42			0.42	0.69			6.1	8.3			10.1	13.6			3.7	5.8			0.5	0.5	
Sudan	0.16				0.35				7.6				17.1				5.0				9.4			
Swaziland	0.03	0.02	0.06	3	0.05	0.05	0.12	3	42.9	25.4	53.0	3	82.9	49.0	102.3	3	16.0	11.4	30.9	3	13.0	5.7	3.1	3
Togo	0.04	0.02	0.03	2	0.09	0.04	0.07	2	14.6	4.6	5.1	2	32.0	10.0	11.2	2	10.7	3.5	3.1	2	7.0	5.6	8.0	2
Uganda	0.01	0.01	0.10		0.03	0.03	0.29		0.7	0.4	3.0		2.1	1.3	8.6		0.6	0.4	3.1		6.7	1.9	3.9	
United Republic of Tanzania	0.11	0.12	0.42	1	0.32	0.34	1.19	1	5.9	4.0	9.5	1	16.9	11.3	27.2	1	6.8	3.1	7.9	1	10.9	8.5	6.7	1
Zambia	0.41	0.03	0.17	1	0.75	0.06	0.32	1	70.4	3.7	13.5	1	130.1	6.8	25.0	1	56.3	3.7	8.9	1	22.8	2.8	9.3	1
Zimbabwe	0.21	0.26	0.33	3	0.14	0.17	0.22	3	29.2	21.9	26.6	3	19.7	14.8	18.0	3	13.5	10.3	12.5	3	7.0	4.2	7.3	3
High-income European countries																								
Austria	2.2	3.3	1.7		2.0	3.0	1.5		276.6	403.2	197.0		250.9	365.7	178.8		22.4	50.4	33.1		2.5	2.4	1.0	
Belgium	2.1				1.8				208.8								38.6				1.7			
Croatia		0.2	0.6			0.3	0.9			36.8	139.4			56.0	212.0			7.7	18.2			3.2	3.6	
Cyprus	0.2	0.2	0.2		0.3	0.2	0.2		399.5	250.7	205.5		435.6	273.3	224.1		48.4	30.1	41.1	1	15.1	4.6	2.1	
Czech Republic		2.4	1.1			2.2	1.0			270.0	85.9			451.8	143.8			57.9	27.6			5.4	1.4	
Denmark	0.8	0.2	0.3		0.7	0.2	0.2		168.4	46.1	46.0		117.5	32.2	32.1		11.6	3.3	10.6		1.0	0.2	0.2	
Estonia		0.0	0.1			0.1	0.1			30.0	66.5			48.1	106.5			10.6	24.4			1.5	1.6	
Finland	7.5	6.9	2.2		6.2	5.7	1.8		1,577.3	1,346.7	416.9		1,297.9	1,108.1	343.0		75.3	112.8	40.8		16.4	8.2	2.0	
France	16.2				14.1				299.9				0.3				25.7				2.8			
Germany	9.1	15.6	7.0		8.5	14.5	6.5		116.5	191.4	85.5		108.1	177.6	79.3		21.8	50.1	29.4		1.2	1.2	0.5	
Greece	4.9	2.7	4.3		5.5	3.0	4.8		7.5	3.7	8.8		8.4	4.1	9.9		0.4	0.3	1.2		0.2	0.1	0.1	

TABLE 2 Continued

	Agricultural expenditure (billions 2005 constant US dollars)				Agricultural expenditure (billions 2005 PPP dollars)				Per capita agricultural expenditure (2005 constant US dollars)				Per capita agricultural expenditure (2005 PPP dollars)				Ratio of agricultural expenditure to agricultural GDP (%)				Share of agriculture in total expenditure (%)			
Region/country	**1980**	**1995**	**2010**	*****	**1980**	**1995**	**2010**	*****	**1980**	**1995**	**2010**	*****	**1980**	**1995**	**2010**	*****	**1980**	**1995**	**2010**	*****	**1980**	**1995**	**2010**	*****
Hungary	9.9	2.4	0.4		9.0	2.2	0.4		613.9	152.9	48.2		952.8	237.3	74.8		46.9	30.7	13.5		10.2	3.6	0.9	
Iceland	0.2	0.2	0.1		0.2	0.2	0.1		1,081.6	711.3	583.6		687.5	452.1	371.0		24.7	15.8	19.0		16.8	7.2	2.3	
Ireland	3.6	1.1	1.1		2.9	0.9	0.9		1,029.2	315.9	240.1		819.2	251.5	191.1		56.3	16.7	56.4		8.0	2.8	0.8	
Italy	10.5	6.7	6.1		9.8	6.2	5.7		146.8	129.1	101.5		136.2	119.7	94.2		12.0	14.4	19.1		1.7	0.9	0.7	
Luxembourg	0.3	0.1	0.2		0.2	0.1	0.2		663.4	347.4	360.3		559.9	293.2	304.1		103.2	58.9	142.9		3.7	1.5	1.0	
Malta	0.0	0.0	0.1		0.0	0.0	0.1		80.8	57.2	123.7		114.8	81.2	175.8		30.7	16.9	39.7		3.3	1.2	1.8	
Netherlands	4.3	1.5	1.7		3.9	1.3	1.5		289.3	92.6	99.6		259.6	83.0	89.4		30.7	8.4	14.1		1.8	0.5	0.5	
Norway	5.2	3.4	2.0		4.7	3.0	1.8			796.0	414.3		911.5	576.5	300.0		85.4	50.3	51.9		8.0	3.0	1.4	
Poland		1.8	3.5			1.6	3.1			44.4	90.2			76.8	156.2			10.7	25.1			3.0	2.0	
Portugal		1.4	0.8			1.6	0.9			135.8	72.4			159.6	85.1			16.4	17.2			2.2	0.8	
Slovakia			1.3	1			1.8	1			234.0	1			332.4	1			66.2	1			4.1	1
Slovenia		0.2	0.2			0.3	0.3			102.3	115.2			134.6	151.5			18.9	26.1			1.8	1.2	
Spain	6.0	4.8	6.1		6.4	5.1	6.4		157.9	120.2	134.2		166.0	126.3	141.1		14.9	13.4	19.4		7.6	1.4	1.1	
Sweden	0.4	0.2	0.8		0.3	0.1	0.7		357.8	174.5	85.0		285.1	139.0	67.7		29.9	18.7	11.3		2.6	0.9	0.4	
Switzerland	7.4	10.0	11.1		6.7	9.1	10.1		364.2	441.2	359.6		839.0	1016.3	933.0		22.6	47.9	56.5		4.9	9.2	7.1	
United Kingdom	5.6	1.3	3.4		5.0	1.2	3.0		116.9	26.7	67.7		101.1	23.1	58.6		26.7	5.1	25.2		1.4	0.2	0.4	
Other high-income countries																								
Australia	1.4	1.7	1.6		1.3	1.6	1.5		94.2	95.0	74.9		88.9	89.6	70.7		5.6	10.3	8.4		1.8	1.2	0.7	
Canada	2.2	3.4	2.4	1	2.2	3.4	2.4	1	88.8	114.9	70.2	1	88.7	114.7	70.1	1	8.9	14.2	13.3	1	2.2	1.9	1.3	1
Israel	0.7	0.6	0.2	3	0.8	0.7	0.3	3	183.8	113.3	31.5	3	222.0	136.8	38.1	3	8.0	6.1	3.3	3	2.0	1.5	0.4	3
Japan	15.1	10.6	16.0	1	12.8	9.0	13.6	1	128.9	84.1	126.0	1	109.7	71.6	107.2	1	17.2	13.7	25.0	1	3.5	1.7	1.7	1
New Zealand	1.2	0.3	0.66	2	1.1	0.3	0.61	2	374.2	74.5	156.8	2	346.3	69.0	145.1	2	19.7	4.5	10.0	2	5.4	1.1	1.4	2
Republic of Korea	1.5	8.6	11.0		2.0	11.1	14.2		41.1	192.0	226.1		53.4	249.2	293.5		5.9	24.7	41.5		5.6	11.6	5.1	
United States of America	18.4	11.9	19.2		18.4	11.9	19.2		80.1	43.9	60.6		80.1	43.9	60.6		10.9	8.2	12.6		1.5	0.6	0.6	

Global Hunger Index

Each year, IFPRI calculates the Global Hunger Index (GHI), which is designed to comprehensively measure and track hunger globally and by country and region. To reflect the multidimensional nature of hunger, the GHI generates one index number from three equally weighted indicators: (1) the percentage of people who are undernourished; (2) the percentage of children younger than age five who are underweight; and (3) the mortality rate of children younger than age five. According to the 2012 GHI, global hunger has declined somewhat since 1990, but remains at a level characterized as "serious" (see specific country scores for 2012 in Table 3). From 1990 to 2012, the greatest improvements in absolute scores took place in Angola, Bangladesh, Ethiopia, Malawi, Nicaragua, Niger, and Vietnam. Hunger worsened, to varying degrees, in Botswana, Burundi, Comoros, Côte d'Ivoire, North Korea, and Swaziland, however. By highlighting successes and failures in hunger reduction and providing insights into the drivers of hunger, the GHI points to the geographic areas where policy action is most needed and suggests policy lessons.

Website: www.ifpri.org/publication/2012-global-hunger-index
Contacts: Klaus von Grebmer or Tolulope Olofinbiyi
Emails: k.vongrebmer@cgiar.org or t.olofinbiyi@cgiar.org

TABLE 3 Global Hunger Index scores (various years), ranked by 2012 country scores

Rank	Country	1990	1996	2001	2012
1	Azerbaijan	–	14.6	7.8	5.0
2	China	11.8	8.9	6.7	5.1
3	Malaysia	9.0	6.7	6.6	5.2
4	Paraguay	7.9	5.8	5.4	5.3
4	Trinidad & Tobago	7.1	7.5	6.2	5.3
6	Gabon	8.4	6.9	7.2	5.4
6	Mauritius	8.0	7.4	6.0	5.4
8	El Salvador	10.1	8.7	5.4	5.7
9	Kyrgyz Republic	–	9.0	9.0	5.8
9	South Africa	6.9	6.5	7.4	5.8
11	Turkmenistan	–	10.0	8.9	6.9
11	Uzbekistan	–	9.0	10.8	6.9
13	Panama	10.1	9.7	8.9	7.0
14	Guyana	14.6	8.9	7.8	7.2
15	Peru	14.5	10.7	9.0	7.4
16	Ecuador	13.5	10.8	8.9	7.5
17	Honduras	13.5	13.2	10.0	7.7
18	Thailand	15.1	11.8	9.2	8.1
19	Suriname	10.3	9.3	10.1	8.5
20	Ghana	21.4	16.3	12.8	8.9
21	Nicaragua	22.4	17.8	12.3	9.1
22	Armenia	–	14.3	11.3	9.2
23	Dominican Republic	14.2	11.8	10.9	10.0
24	Swaziland	9.3	12.6	12.9	10.9
25	Mauritania	22.6	16.7	16.6	11.1
26	Vietnam	25.6	21.4	15.5	11.2
27	Congo, Rep.	23.6	24.1	15.7	11.4
28	Mongolia	16.5	17.5	14.8	11.7
29	Lesotho	12.6	13.6	13.9	11.9
30	Indonesia	18.5	15.4	14.2	12.0
31	Philippines	19.9	17.6	14.2	12.2
32	Bolivia	16.9	14.3	12.3	12.3
33	Guatemala	15.2	15.8	15.1	12.7
34	Namibia	20.3	19.1	16.3	13.2
35	Botswana	13.4	15.4	15.7	13.7
35	Senegal	18.3	19.6	19.2	13.7
37	Sri Lanka	20.8	18.4	15.2	14.4
38	Benin	21.3	20.1	16.8	14.6
39	Gambia, The	16.2	20.1	16.3	15.6
40	Nigeria	24.1	20.9	18.2	15.7
41	Tajikistan	–	24.1	24.6	15.8
42	Uganda	18.7	20.3	17.3	16.1

GHI Severity Scale

≥ 30.0	Extremely alarming	5.0–9.9	Moderate
20.0–29.9	Alarming	< 5.0	Low
10.0–19.9	Serious	–	No data

Notes: Countries with a 2012 GHI score of less than five are not included in the ranking but are shown in Table 4. Countries that have identical 2012 GHI scores are given the same ranking (for example, Paraguay and Trinidad & Tobago both rank fourth). The following countries could not be included owing to lack of data: Afghanistan, Bahrain, Bhutan, Democratic Republic of Congo, Iraq, Myanmar, Oman, Papua New Guinea, Qatar, and Somalia. Data for calculating the 1990 GHI are from 1988–1992; for the 1996 GHI, from 1994–1998; for the 2001 GHI, from 1999–2003; and for the 2012 GHI, from 2005–2010.

TABLE 3 Continued

Rank	Country	1990	1996	2001	2012
43	Mali	27.8	26.3	23.0	16.2
44	Guinea	22.4	20.0	21.6	16.6
45	Malawi	29.9	27.5	22.5	16.7
46	Burkina Faso	23.5	22.4	21.8	17.2
47	Zimbabwe	18.6	22.3	21.3	17.3
48	Cameroon	21.6	22.2	19.0	17.4
49	Côte d'Ivoire	16.5	17.8	16.6	18.2
50	Guinea-Bissau	20.7	20.8	21.4	18.4
51	Liberia	22.7	25.2	25.0	18.9
52	North Korea	15.7	20.1	20.1	19.0
52	Togo	26.4	22.0	23.3	19.0
54	Kenya	20.7	20.8	20.4	19.3
54	Tanzania	23.2	28.0	25.9	19.3
56	Cambodia	31.8	31.5	26.0	19.6
57	Lao PDR	28.6	25.2	23.6	19.7
57	Pakistan	25.5	21.8	21.7	19.7
57	Rwanda	28.2	32.7	25.6	19.7
60	Nepal	26.9	24.4	23.0	20.3
61	Sudan	28.7	24.5	25.9	21.5
62	Djibouti	30.8	25.7	25.3	21.7
63	Niger	36.4	35.9	30.5	22.3
64	Madagascar	24.1	23.8	24.9	22.5
65	India	30.3	22.6	24.2	22.9
66	Mozambique	35.5	30.7	28.8	23.3
66	Zambia	24.8	25.0	27.2	23.3
68	Bangladesh	37.9	36.1	27.8	24.0
69	Angola	41.9	39.9	33.0	24.1
70	Yemen, Rep.	29.0	27.6	27.9	24.3
71	Sierra Leone	32.7	30.1	30.1	24.7
72	Comoros	22.2	26.9	29.7	25.8
73	Central African Rep.	27.4	28.4	27.4	27.3
73	Timor-Leste	–	–	26.1	27.3
75	Chad	39.3	35.6	30.4	28.3
76	Ethiopia	42.2	38.6	34.5	28.7
77	Haiti	33.9	32.2	25.8	30.8
78	Eritrea	–	37.8	37.8	34.4
79	Burundi	31.6	35.9	38.0	37.1

TABLE 4 Countries with 2012 Global Hunger Index scores of less than 5

Country	1990	1996	2001	2012
Albania	8.5	5.2	8.2	<5
Algeria	6.7	7.3	6.0	<5
Argentina	<5	<5	<5	<5
Belarus	–	<5	<5	<5
Bosnia & Herzegovina	–	<5	<5	<5
Brazil	7.4	6.4	5.4	<5
Bulgaria	<5	<5	<5	<5
Chile	<5	<5	<5	<5
Colombia	9.2	6.8	5.8	<5
Costa Rica	<5	<5	<5	<5
Croatia	–	<5	<5	<5
Cuba	<5	6.5	<5	<5
Egypt, Arab Rep.	8.0	6.7	5.3	<5
Estonia	–	<5	<5	<5
Fiji	6.3	<5	<5	<5
Georgia	–	8.7	6.0	<5
Iran, Islamic Rep.	8.8	7.3	5.1	<5
Jamaica	6.7	5.0	<5	<5
Jordan	<5	<5	<5	<5
Kazakhstan	–	<5	5.4	<5
Kuwait	9.1	<5	<5	<5
Latvia	-	<5	<5	<5
Lebanon	<5	<5	<5	<5
Libya	<5	<5	<5	<5
Lithuania	-	<5	<5	<5
Macedonia, FYR	-	<5	<5	<5
Mexico	7.9	5.4	<5	<5
Moldova	-	5.7	5.2	<5
Montenegro	-	-	-	<5
Morocco	7.6	6.8	6.2	<5
Romania	<5	<5	<5	<5
Russian Federation	-	<5	<5	<5
Saudi Arabia	6.3	6.2	<5	<5

Note: Countries with 2012 GHI scores of less than five are not ranked because differences among their 2012 scores are minimal.

TABLE 4 Continued

Country	1990	1996	2001	2012
Serbia	-	-	-	<5
Slovak Republic	-	<5	<5	<5
Syrian Arab Republic	6.7	5.7	5.4	<5
Tunisia	<5	<5	<5	<5

TABLE 4 Continued

Country	1990	1996	2001	2012
Turkey	5.7	5.3	<5	<5
Ukraine	-	<5	<5	<5
Uruguay	<5	<5	<5	<5
Venezuela, RB	6.7	7.1	6.4	<5

New Food Security Indexes

ALEXANDER J. STEIN

In the wake of the recent food price crises, attention has turned to food security and how it is measured. To better understand levels of and trends in hunger, at least at an aggregate level, players in industry, the nonprofit sector, and government launched a number of indexes in 2012, including the Global Food Security Index, the Rice Bowl Index, the Agricultural Transformation Index, and the Nutrition Barometer.[1] These efforts add to recent work such as the Hunger Reduction Commitment Index and the Food Security Risk Index.[2]

Other measures of national food security have been around for some time. Examples include reporting by the Food and Agriculture Organization of the United Nations (FAO) of the number of undernourished people, the US Department of Agriculture's analysis of food availability within countries, and the quantification of the burden of undernutrition and related risk factors by the World Health Organization (WHO).[3] IFPRI's own Food Security Vulnerability Indices take into account trade flows to detect countries' exposure to world agricultural markets, and its Global Hunger Index combines key outcomes of food and nutrition security to show the status of the hunger problem.[4]

Such indexes are useful, but they have their limitations.[5] New indexes often simply reorganize existing data. Although their designs and methods may be sound, little is gained if the quality, accuracy, and timeliness of the underlying data are poor. Generating new data or updating data more often could help improve the measurement of food security. Examples of such efforts are the Women's Empowerment in Agriculture Index, IFPRI's Agricultural Science and Technology Indicators (ASTI), and the Living Standards Measurement Study of the World Bank.[6] Crucially, the FAO reviewed its methodology for estimating undernourishment and included an additional set of indicators in its assessment of the world's undernourished people.[7]

Other problems relate to how indexes are constructed. Distortions can arise if an index integrates several indicators that are driven by the same underlying factors, such as income or governance. Similarly, an index that relies on indirect indicators for food security—such as prices—can be ambivalent. For instance, low agricultural prices may strengthen consumers' food security but harm that of producers. In contrast, the Global Hunger Index and WHO's measure of the burden of disease include "output" indicators—that is, they measure the actual consequences of food insecurity such as undernourishment and poor health. Whereas input indicators may be useful early warning tools if they capture changes in a timely manner, output indicators can help document the status of food insecurity and measure progress in the fight against hunger, thereby serving as tools for advocacy and accountability.

The proliferation of food security indexes demonstrates the widespread interest in understanding the size of the hunger problem. To improve the measurement of food security, however, efforts and resources should be concentrated not on producing "index inflation," but on supporting fewer indexes that rely on sound conceptual frameworks and that integrate new and timely data.

Alexander J. Stein is a research coordinator in the Director General's Office of the International Food Policy Research Institute in Washington, DC.

Food Policy Research Capacity Indicators

The motivation to measure food policy research capacity stems from a pressing need to understand what a country's primary constraints are in undertaking food policy research and using it effectively in the policy process. With such understanding, capacity-development interventions can become easier to design and capacity-building programs can be more effective. Food policy research is defined here as any socioeconomic or policy-related research in the food, agriculture, and natural-resource sectors. IFPRI's quantitative assessment of selected developing countries' capacity to conduct food policy research and analysis was presented for the first time in the *2011 Global Food Policy Report*.[1] This was a preliminary attempt to identify and document capacity indicators. The methodology has been refined and the data has been updated since then. In particular, the database has been expanded to include 30 countries and a broader range of organizations within the countries; we are continuing to refine our selection of indicators and to expand our data collection in other countries.

Two country-level food policy research capacity indicators are presented here. The first indicator is the number of full-time equivalent analysts or researchers with PhDs or their equivalent per one million rural citizens. This number is indicative of a country's investment in policy research activities. Although financial and physical resources are not explicitly included in this measure, it is assumed that the number of researchers acts as a proxy for these other food policy research inputs. This indicator includes staff at government ministries, higher-education institutes, and research organizations that undertake food policy research as defined above. Staff members with a master's degree are valued at half of a PhD, and those with a bachelor's degree are valued at a quarter of a PhD. The number of staff is also scaled by the proportion of time spent on food policy research, which is dependent on the type of institution.

The second indicator is the number of food policy–relevant journal articles published internationally within the last five years per full-time PhD-equivalent researcher. This input-output ratio measure is indicative of the efficiency of the policy research environment. The number of publications was determined from searches in the EconLit and Web of Science databases for journal articles related to food policy and authored by experts who were counted in the assessment of the first indicator. Earlier attempts to quantify and collect comparable data on other policy research outputs, such as policy briefs, interactions with government ministries, or conference contributions presented numerous challenges. For this reason, and because international publications guarantee a minimum and comparable level of quality, this indicator was chosen.

It is hoped that these indicators will sensitize domestic policymakers to the importance of various components of food policy research, allow donors to set priorities for interventions across countries, and identify the specific areas of capacity in which to invest in order to maximize social returns. Additionally, the indicators will facilitate cross-country comparisons of food policy research capacity, in the same way that the UN's Human Development Index or IFPRI's Global Hunger Index has done with indicators of social development. Once fully developed, these food policy research capacity indicators could be used to undertake an analysis of the role and impact of food policy research on food and agricultural systems.

Organizational capacity and the enabling environment in which individuals and organizations interact are increasingly recognized as essential to ensuring inputs are effectively and efficiently transformed into implementable policies. Future work will combine the current indicators with measures of organizational and system-level capacity to provide a more comprehensive picture of the effectiveness of a country's food policy research capacity.

Contacts: Suresh Babu or Paul Dorosh
Emails: s.babu@cgiar.org or p.dorosh@cgiar.org

TABLE 5 Food policy research capacity indicators, 2011

Country	Analysts/ researchers (head count), 2011	Full-time equivalent analysts/ researchers with PhD equivalent, 2011	International publications produced, 2007–2011	Full-time equivalent analysts/ researchers with PhD equivalent per million rural population, 2011	Publications per full-time equivalent analysts/ researchers with PhD equivalent, 2007–2011
Afghanistan	43	2.975	0	0.110	0.000
Bangladesh	66	22.900	39	0.212	1.703
Benin	38	4.300	18	0.858	4.186
Burundi*	39	5.125	0	0.671	0.000
Colombia	85	6.450	24	0.556	3.721
Ethiopia	74	22.350	28	0.318	1.253
Ghana	95	20.275	46	1.687	2.269
Guatemala	45	11.900	5	1.606	0.420
Honduras*	33	6.125	4	1.651	0.653
Kenya	155	31.600	55	0.999	1.741
Laos	9	1.750	1	0.423	0.571
Liberia	34	3.075	0	1.437	0.000
Madagascar*	187	11.525	7	0.802	0.607
Malawi	50	10.525	6	0.812	0.570
Mali	60	10.050	0	0.975	0.000
Mozambique	37	4.700	16	0.286	3.404
Nepal*	27	3.650	4	0.144	1.096
Niger*	29	8.825	3	0.669	0.340
Nigeria	349	77.400	22	0.946	0.284
Peru*	54	7.150	16	1.069	2.238
Rwanda	37	3.200	1	0.362	0.313
Senegal	71	9.300	5	1.268	0.538
South Africa	198	50.325	217	2.617	4.312
Swaziland	32	2.850	3	3.391	1.053
Tanzania	91	20.750	12	0.613	0.578
Togo	81	6.825	11	1.789	1.612
Uganda	34	10.925	14	0.375	1.281
Vietnam	175	32.525	5	0.537	0.154
Zambia*	29	5.300	8	0.647	1.509
Zimbabwe*	42	8.875	16	1.134	1.803

* The number of analysts or researchers or the number of institutions surveyed changed between 2010 and 2011.

Total and Partial Factor Productivity

Total factor productivity (TFP) is the ratio of total output (crop and livestock products) to total production inputs (land, labor, capital, and materials). An increase in TFP implies that more output is being produced from a constant amount of resources used in the production process. In the long run, TFP is the main driver of growth in agriculture and can be affected by policies and investment. Partial factor productivity (PFP) measures, such as labor and land productivity, are often used to measure agricultural-production performance because they are easy to estimate. These measures of productivity normally show higher rates of growth than TFP because growth in land and labor productivity could result from more intensive use of inputs, including fertilizer and machinery, rather than TFP increase. If productivity increases without the addition of more inputs, then the only source of growth is TFP.

Table 6 presents estimates of IFPRI's TFP and PFP measures for developing countries for three sub-periods between 1981 and 2009 (1981–1990, 1991–2000, and 2001–2009). These TFP and PFP estimates were generated using data from the Food and Agriculture Organization of the United Nations (FAO) on outputs and inputs. The output values are the FAO-constructed gross agricultural outputs, measured in constant 2004–2006 US dollars and smoothed using the Hodrick-Prescott filter. Each output value is a composite of 190 crop and livestock commodities aggregated using a constant set of global average prices from 2004–2006. Inputs include agricultural land, measured by the sum, in hectares, of cropland and permanent pasture; labor, measured by the number of economically active persons in agriculture; livestock, measured by the number of animals in cattle equivalents; machinery, measured by the total amount of horsepower available from four-wheel tractors, pedestrian-operated tractors, and combine-threshers in use; and fertilizer, measured by tons of fertilizer nutrients used.[2] This dataset of outputs and inputs was checked and cleaned using different statistical techniques.

TFP estimates were obtained using Data Envelopment Analysis (DEA) techniques. These techniques have been extensively used because they make TFPs easy to compute, do not involve restrictive assumptions such as specific production functions (used to calculate productivity) or constant input shares across countries, and do not require information about input prices or assumptions regarding economic behavior, such as cost minimization or profit maximization. On the other hand, DEA productivity estimates are sensitive to data noise and outliers and can suffer from the problem of "unusual" weights that are higher or lower than expected when aggregating inputs to measure TFP. Given these limitations, outlier detection methods were used to determine influential observations in the dataset and input weights were allowed to vary only within a certain range of expected values because specific lower and upper bounds were imposed for each input in different regions.[3] Results are also affected by data characteristics and quality issues. In particular, the data series on fertilizer and machinery show high volatility and could result in high variability of TFP estimates for some countries.

Contact: Alejandro Nin-Pratt
Email: a.ninpratt@cgiar.org

TABLE 6 Levels of land and labor productivity and average annual growth of agricultural output and total factor productivity (TFP) in developing countries, 1980–2009

Region/country	1980		1981–1990		1990		1991–2000		2000		2001–2009		2009	
	Land productivity	Labor productivity	Output growth	TFP growth	Land productivity	Labor productivity	Output growth	TFP growth	Land productivity	Labor productivity	Output growth	TFP growth	Land productivity	Labor productivity
	In constant 2004–2006 US $		%		In constant 2004–2006 US $		%		In constant 2004–2006 US $		%		In constant 2004–2006 US $	
Africa south of Sahara	163	1,219	2.86	0.98	182	1,315	3.11	1.82	192	1,661	2.63	1.65	219	2,057
Angola	59	312	0.82	−0.48	63	258	4.77	2.99	64	315	7.32	4.46	70	451
Benin	255	712	4.84	−1.04	275	820	6.11	3.28	264	1,175	2.83	3.69	242	1,318
Botswana	62	927	1.31	0.93	70	1,055	−0.58	−5.19	78	728	3.09	2.39	87	862
Burkina Faso	144	212	6.03	0.12	252	294	4.37	0.23	400	338	3.22	0.08	395	335
Burundi	164	442	2.56	−0.81	169	412	−0.87	−0.51	159	348	−1.14	−4.21	160	235
Cameroon	288	690	2.11	0.74	347	700	3.16	1.67	345	847	3.14	2.79	330	1,094
Central African Republic	127	514	2.02	0.23	160	521	3.51	2.11	169	643	2.12	1.57	187	742
Chad	41	516	2.48	−1.44	52	456	3.63	−0.08	71	509	2.19	0.77	73	511
Congo	19	447	1.54	−0.04	24	462	2.87	0.88	28	546	3.46	4.36	27	713
Congo, DR	70	439	3.11	1.11	82	474	−1.75	−2.00	64	320	−0.29	−0.04	70	262
Côte d'Ivoire	77	1,351	3.43	0.69	89	1,422	3.47	1.16	109	1,823	1.40	3.10	110	2,160
Ethiopia	327	311	0.89	−2.60	332	248	2.61	−1.42	355	230	4.89	2.66	408	275
Gabon	26	813	2.36	0.95	42	991	1.75	1.62	43	1,179	1.28	−1.33	53	1,470
Ghana	100	585	4.58	4.17	153	696	5.70	2.50	125	907	4.17	1.22	172	1,058
Guinea	89	421	2.52	0.49	98	435	3.72	−0.37	148	447	3.22	1.60	138	525
Guinea-Bissau	100	360	3.57	1.80	125	437	3.52	0.85	131	534	1.94	−0.10	169	564
Kenya	232	486	3.98	−0.04	309	505	1.46	−0.52	327	426	3.40	2.06	339	477
Liberia	66	582	−1.04	−0.60	52	501	3.28	2.42	61	530	1.05	−1.00	83	469
Madagascar	118	670	1.62	1.06	119	624	0.52	−0.06	106	505	2.63	1.50	100	476
Malawi	198	356	1.29	−1.19	204	302	5.64	4.65	184	450	4.98	2.52	246	570
Mali	105	634	2.96	−0.41	109	759	3.00	1.23	131	838	5.02	3.25	165	1,038
Mauritania	36	665	1.56	−2.38	44	762	1.71	1.02	52	689	1.98	−0.54	56	645
Mozambique	47	230	−1.02	−0.31	36	202	5.28	3.00	56	246	1.58	0.30	57	236
Namibia	37	2,159	0.08	−0.21	41	1,747	0.14	−1.92	45	1,531	0.90	−1.50	54	1,580
Niger	85	569	0.89	−0.98	80	483	5.25	1.90	102	584	6.26	4.14	86	762
Nigeria	155	759	6.02	3.17	188	1,368	4.91	3.61	183	2,249	1.99	1.54	239	2,718
Rwanda	166	440	1.82	−0.20	185	403	0.67	0.06	147	367	4.61	−0.26	233	422
Senegal	199	423	1.76	0.03	200	403	1.53	−1.20	262	367	3.62	1.25	356	397
Sierra Leone	99	411	1.54	−0.18	113	395	−1.22	1.25	99	364	5.97	3.34	119	485
Somalia	100	685	0.73	0.38	106	738	0.69	1.92	99	720	0.81	−1.12	105	662
South Africa	173	5,132	0.92	1.21	167	5,594	1.30	4.27	147	6,928	2.16	3.80	151	10,191

Note: Land productivity is agricultural gross production per hectare of agricultural land; labor productivity is agricultural gross production per economically active person in agriculture. Both types of agricultural gross production are measured in constant 2004–2006 US dollars.

Source: Food and Agriculture Organization of the United Nations, FAOSTAT database, faostat.fao.org, accessed January 23, 2013.

TABLE 6 Continued

	1980		1981–1990		1990		1991–2000		2000		2001–2009		2009	
	Land productivity	Labor productivity	Output growth	TFP growth	Land productivity	Labor productivity	Output growth	TFP growth	Land productivity	Labor productivity	Output growth	TFP growth	Land productivity	Labor productivity
Region/country	In constant 2004–2006 US $		%		In constant 2004–2006 US $		%		In constant 2004–2006 US $		%		In constant 2004–2006 US $	
Sudan	119	827	1.05	−0.09	134	829	4.97	3.92	155	1,115	1.82	−1.13	184	1,162
Tanzania	225	392	2.21	0.71	226	361	1.84	−0.63	293	337	4.53	1.68	292	413
Togo	127	497	3.11	−1.42	169	518	3.76	2.49	180	616	1.39	−0.79	205	607
Uganda	169	601	2.43	2.26	211	566	2.87	−0.04	284	595	1.34	−0.15	305	526
Zambia	112	368	3.45	0.42	134	345	1.45	1.27	122	328	4.18	2.70	125	406
Zimbabwe	306	645	1.73	0.27	328	533	2.15	0.99	298	577	−2.47	1.65	263	485
Latin America and Caribbean	326	5,425	2.35	0.94	368	5,799	3.15	1.79	394	7,965	3.42	3.47	424	11,533
Argentina	234	17,053	0.65	0.32	234	16,338	3.02	−0.16	251	22,000	2.28	0.94	279	27,753
Bolivia	109	1,246	2.71	−0.39	112	1,377	3.92	2.61	137	1,542	3.06	−2.41	167	1,633
Brazil	354	2,729	3.51	2.94	424	4,476	3.65	1.93	465	6,761	4.64	5.13	489	11,951
Chile	236	4,078	3.43	0.74	257	4,666	3.86	0.55	354	6,612	2.23	2.88	328	8,016
Colombia	279	2,205	2.35	2.04	304	2,830	1.80	4.46	309	3,244	2.99	2.65	343	4,157
Costa Rica	467	3,943	3.92	2.70	687	5,153	3.58	3.65	575	6,892	2.64	2.54	597	8,761
Dominican Republic	444	2,888	0.99	−0.22	526	2,909	0.40	1.73	540	3,438	3.25	3.00	618	5,372
Ecuador	349	2,466	3.49	1.44	387	3,054	4.35	−0.66	481	4,311	3.65	3.40	500	5,848
El Salvador	533	1,415	−0.43	−0.77	571	1,305	1.54	1.76	536	1,506	2.06	1.94	590	2,002
Guatemala	426	1,299	2.11	0.75	422	1,301	3.87	3.24	420	1,895	4.06	1.98	492	2,000
Haiti	319	584	−0.38	−0.65	341	522	−0.23	−4.64	458	456	1.42	0.60	408	458
Honduras	342	1,571	1.19	−0.36	388	1,706	1.69	−0.57	446	1,843	4.24	3.06	529	2,943
Jamaica	579	1,415	0.65	−0.99	709	1,625	2.28	2.36	784	2,251	−0.01	5.52	604	2,569
Mexico	281	2,413	1.72	−2.96	337	2,663	2.87	3.21	342	3,444	1.90	2.86	355	4,410
Nicaragua	421	2,290	−3.01	−3.49	320	1,645	4.27	4.02	400	2,501	4.09	2.59	337	3,941
Panama	457	3,741	0.94	−0.29	409	3,175	0.55	−1.62	452	3,212	1.16	2.58	490	3,678
Paraguay	209	3,200	4.88	−0.60	227	4,411	1.75	−7.61	278	4,222	3.51	−0.41	320	5,014
Peru	309	1,348	1.84	0.42	311	1,274	4.98	3.48	321	1,714	4.35	3.53	358	2,290
Trinidad and Tobago	1,072	3,167	−1.11	1.69	1,459	2,600	0.95	0.56	1,899	2,855	1.89	4.42	2,701	3,593
Uruguay	359	10,305	1.12	1.02	353	11,941	2.44	0.41	363	14,162	3.58	3.60	415	20,430
Venezuela	310	4,538	2.73	−0.55	377	4,922	2.61	3.81	392	6,805	1.91	1.62	444	8,909
Asia	494	594	4.08	0.66	607	772	3.99	2.07	704	1,022	3.48	2.59	773	1,522
Bangladesh	801	366	2.02	1.53	955	357	2.83	2.01	1,280	457	4.19	1.32	1,414	651
Cambodia	221	272	6.66	3.49	313	385	4.24	−0.78	396	454	7.53	2.64	505	717
China	358	489	4.92	1.54	432	717	5.32	3.10	526	1,073	3.41	3.20	576	1,761
India	690	516	3.46	−0.32	877	616	2.55	0.79	1,029	695	2.85	1.99	1,135	804

TABLE 6 Continued

Region/country	1980		1981–1990		1990		1991–2000		2000		2001–2009		2009	
	Land productivity	Labor productivity	Output growth	TFP growth	Land productivity	Labor productivity	Output growth	TFP growth	Land productivity	Labor productivity	Output growth	TFP growth	Land productivity	Labor productivity
	In constant 2004–2006 US $		%		In constant 2004–2006 US $		%		In constant 2004–2006 US $		%		In constant 2004–2006 US $	
Indonesia	221	576	4.84	−0.18	324	706	2.38	0.30	394	791	4.83	1.40	429	1,183
Laos	376	392	3.68	−0.32	417	441	5.28	0.48	684	587	5.10	2.91	555	740
Malaysia	330	2,364	4.49	1.16	344	3,887	2.82	0.13	409	5,367	4.08	1.39	499	8,666
Mongolia	44	3,566	1.02	4.49	44	3,736	−0.88	4.26	38	3,521	0.94	0.36	56	4,019
Myanmar	469	475	1.37	−0.80	554	445	4.90	2.46	666	606	7.46	6.72	735	1,065
Nepal	649	334	4.65	0.26	806	431	2.98	−2.36	1,122	443	2.80	3.02	1,124	421
Pakistan	708	997	4.67	−0.79	980	1,396	3.30	0.79	1,133	1,552	3.11	−0.15	1,451	1,594
Philippines	573	1,140	1.69	−0.49	745	1,120	2.57	0.21	854	1,263	3.45	2.36	930	1,593
Sri Lanka	695	655	−0.05	0.49	704	566	1.17	1.61	616	630	2.13	1.18	630	686
Thailand	445	844	2.77	−0.61	617	880	2.27	3.29	692	1,167	2.81	1.76	739	1,542
Viet Nam	752	383	4.54	2.18	850	470	5.91	0.48	1,146	729	4.47	2.09	1,203	957
Middle East and North Africa	519	2,828	3.70	1.43	562	3,340	2.87	2.23	640	4,472	2.37	1.81	706	5,570
Algeria	110	1,205	4.48	3.33	163	1,599	2.70	2.63	169	1,464	4.53	3.52	198	1,876
Egypt	1,429	1,105	4.13	1.15	1,700	1,565	4.76	2.20	1,837	2,661	3.49	1.59	1,975	3,464
Iran	219	1,843	5.24	0.73	283	2,594	4.34	2.88	330	3,468	2.57	1.13	426	3,842
Iraq	292	2,559	2.46	−0.02	297	4,203	0.88	4.16	344	5,361	−1.62	−2.31	330	5,580
Jordan	253	3,714	6.90	3.98	330	5,391	3.26	0.40	405	6,395	4.06	5.22	457	9,462
Kuwait	222	5,885	4.61	4.27	481	8,012	7.24	12.20	666	13,477	3.75	2.05	781	14,734
Lebanon	328	5,192	5.72	6.96	500	15,859	1.17	0.73	629	25,480	0.70	3.11	586	43,387
Libya	101	3,213	2.78	2.35	124	6,253	2.88	5.11	108	10,221	0.76	0.67	125	15,235
Morocco	202	989	5.25	4.14	221	1,564	1.15	−0.56	261	1,697	3.98	3.63	241	2,669
Oman	139	716	4.88	−2.58	206	717	4.75	2.95	252	995	1.41	0.27	222	1,060
Qatar	388	4,311	8.21	3.42	606	4,059	4.28	6.21	622	10,727	4.04	3.48	888	7,561
Saudi Arabia	25	747	11.94	4.22	34	2,493	1.43	−1.33	31	4,208	2.84	2.52	33	6,624
Syria	166	4,551	1.94	−2.50	243	3,893	4.26	3.57	276	5,048	1.54	−0.78	308	4,845
Tunisia	197	2,341	3.89	2.01	248	3,617	2.17	0.43	289	3,861	2.59	1.65	307	4,580
Turkey	524	2,528	2.43	1.15	513	2,548	1.54	2.43	509	3,368	1.46	2.54	504	4,285
United Arab Emirates	341	4,031	9.34	3.80	588	3,354	14.07	8.88	984	10,443	−1.96	−2.64	684	5,287
Yemen	81	544	2.92	0.54	99	581	3.60	2.05	116	591	5.00	3.23	141	758

Knowledge Is Power: Open Access in 2012

GWENDOLYN STANSBURY AND LUZ MARINA ALVARÉ

"Knowledge is power. Information is liberating."

—Kofi Annan

In the 20th century, technological developments made it possible for vast amounts of information to be found, used, and exchanged quickly and freely, but a great deal of valuable research and data still remains inaccessible. Information can only be liberating if it is liberated.

The open access and open data movements seek to do just that by opening access to scholarly publications, research, and data held by publishers, institutions, public agencies, and others—so that anyone is free to use, reuse, and redistribute the information, subject to attribution. There have been great strides in recent years, as documented by sites such as ROARMAP (Registry of Open Access Repositories Mandatory Archiving Policies),[1] COAR (Confederation of Open Access Repositories),[2] and data.gov (which lists US and international open data sites).

In 2012, a flurry of activity brought the issue of open access and open data into the global spotlight, especially with regard to publically funded research and the reusability of that work.[3]

- Argentina, Brazil, Colombia, Mexico, Chile, Ecuador, Peru, Venezuela, and El Salvador agreed to develop a Latin American network of institutional repositories for scientific and technological research.
- The European Commission announced an open access mandate for 2014–2020 research funding and created the OpenAIRE repository.[4]
- India's Union Cabinet approved a National Data Sharing and Accessibility Policy and launched an open data website.[5]
- The global Open Government Partnership—launched in September 2011 by Brazil, Indonesia, Mexico, Norway, Philippines, South Africa, United Kingdom, and the United States—now includes more than 50 member countries.
- The UK Department for International Development established an open and enhanced access policy for the research it funds.[6] The Research Councils UK updated its open access policy, with an emphasis on maximizing the reusability of research outputs.[7]
- The Wellcome Trust announced it will begin enforcing compliance with its open access policy for the research that it funds, with immediate effect.[8]
- The White House received and must respond to an official petition from US constituents to implement open access policies for all federal agencies that fund scientific research.[9]
- The World Bank enacted its Open Access Policy for research and knowledge.[10]

Freely available research and data empowers people and institutions to make informed decisions. It can inspire innovation, improve the delivery of public services, hold governments accountable, and spur social and economic growth. For example, the open access Food Security Portal facilitated by IFPRI provides comprehensive, country-by-country information on commodity prices and food policy developments to help build countries' resilience to food price shocks and contribute to their food security in the long term.

The hope is that the world is fast approaching the open access tipping point. That will be the moment that knowledge will truly be power.

For more on IFPRI's commitment to providing openly accessible knowledge, see its Open Access Statement online.[11]

Gwendolyn Stansbury is the director of the Communications and Knowledge Management Division and **Luz Marina Alvaré** is the head of the Knowledge Management Unit at the International Food Policy Research Institute in Washington, DC.

NOTES

CHAPTER 1

[1] US Office of Management and Budget, "Department of State and Other International Programs," in *Budget of the United States Government, Fiscal Year 2012*, www.whitehouse.gov/sites/default/files/omb/budget/fy2012/assets/state.pdf; Deutsche Gesellschaft für Internationale Zusammenarbeit (GIZ), Securing Food: Present and Future Approaches to Food Security, December 11, 2012, www.giz.de/en/mediacenter/9774.html; DFID, Hunger Event: Lasting Legacy for Children around the World, August 12, 2012, www.dfid.gov.uk/News/Latest-news/2012/Hunger-event-news-story/; World Bank, Agriculture and Rural Development, updated September 2012, http://go.worldbank.org/BR4VX141Q0.

[2] Climate Central, *Book It: 2012, the Hottest US Year on Record* (Princeton, NJ, US, 2012).

[3] United States Department of Agriculture, Economic Research Service, US Drought 2012: Farm and Food Impacts, www.ers.usda.gov/topics/in-the-news/us-drought-2012-farm-and-food-impacts.aspx, accessed December 13, 2012.

[4] FAO, *Executive Brief: The Sahel Crisis 2012*, June 21 (Rome, 2012).

[5] FAO, *Global Food Price Monitor*, September 12 (Rome, 2012).

[6] World Food Programme, *Global Food Security Update: Tracking Food Security Trends in Vulnerable Countries*, Issue 8 (October) (Rome, 2012).

[7] FAO, *Crop Prospects and Food Situation*, No. 3 (October) (Rome, 2012).

[8] X. Diao, A. Kennedy, A. Mabiso, and A. Pradesha, *Economywide Impact of Maize Export Bans on Agricultural Growth and Household Welfare in Tanzania*, IFPRI Discussion Paper (Washington, DC: International Food Policy Research Institute, forthcoming).

[9] United Nations High Commissioner for Refugees (UNHCR), UNHCR Global Appeal 2013 Update: Mali Situation, www.unhcr.org/50a9f82316.html, accessed December 21, 2012.

[10] UNHCR, UNHCR Global Appeal 2013 Update: Democratic Republic of the Congo, www.unhcr.org/50a9f81f0.html, accessed December 20, 2012.

[11] USAID, *Somalia Food Security Outlook*, April to September 2012, www.fews.net/docs/Publications/Somalia_OL_2012_04_final.pdf, accessed December 20, 2012.

[12] World Food Programme, Syria: Overview, www.wfp.org/countries/syria/overview, accessed December 17, 2012.

[13] United Nations General Assembly, "The Future We Want," www.uncsd2012.org/content/documents/727The%20Future%20We%20Want%2019%20June%201230pm.pdf, accessed January 18, 2013.

[14] European Commission, "Innovating for Sustainable Growth: A Bioeconomy for Europe," February 13 (Brussels, 2012), http://ec.europa.eu/research/bioeconomy/pdf/201202_innovating_sustainable_growth.pdf.

[15] World Bank, *World Development Report 2012* (Washington, DC, 2012); FAO, *State of Food and Agriculture 2010–2011* (Rome, 2011).

[16] USAID, IFPRI, and Oxford Poverty and Human Development Initiative (OPHI), Women's Empowerment in Agriculture Index, www.ifpri.org/publication/womens-empowerment-agriculture-index.

[17] IFAD, Youth in Agriculture, Special session of the Farmers' Forum Global Meeting, February 18, 2012, Rome, www.ifad.org/farmer/2012/youth/snthnesis.htm.

[18] D. Orden, D. Blandford, and T. Josling, *WTO Disciplines on Agricultural Support: Seeking a Fair Basis for Trade* (Cambridge, UK: Cambridge University Press, 2011).

[19] World Bank, World Development Indicators 2012, http://data.worldbank.org, accessed January 10, 2013.

[20] FAO, *Rice Market Monitor* 15, no. 4 (November) (Rome, 2012).

[21] Gulf Cooperation Council member countries are Bahrain, Kuwait, Oman, Qatar, Saudi Arabia, and the United Arab Emirates.

CHAPTER 2

[1] S. Zhao, Y. Sheng, and E. M. Gray, "Measuring Productivity of the Australian Broadacre and Dairy Industries: Concepts, Methodology and Data" and S. L. Wang, D. Schimmelpfennig, and K. Fuglie, "Is Agricultural Productivity Growth Slowing in Western Europe?" in *Productivity Growth in Agriculture: An International Perspective*, ed. K. Fuglie, S. L. Wang, and V. Eldon Ball (Oxfordshire, England: CAB International, 2012).

[2] K. O. Fuglie and S. L. Wang, "New Evidence Points to Robust but Uneven Productivity Growth in Global Agriculture," *Amber Waves* 10 (September 2012); A. Nin-Pratt, *Trends in Global Agriculture*, IFPRI Discussion Paper (Washington, DC: IFPRI, forthcoming).

[3] K. Fuglie, "Productivity Growth and Technology Capital in the Global Agricultural Economy," in *Productivity Growth in Agriculture: An International Perspective*, ed. K. Fuglie, S. L. Wang, and V. Eldon Ball (Oxfordshire, England: CAB International, 2012).

[4] A. Nin-Pratt, *Trends in Global Agriculture*.

[5] K. Fuglie, S. L. Wang, and V. Eldon Ball, ed., *Productivity Growth in Agriculture: An International Perspective* (Oxfordshire, England: CAB International, 2012).

[6] A. Nin-Pratt, "Technological Change and the Transformation of Global Agriculture: From Biotechnology and Gene Revolution to Nano Revolution," in *Nanotechnology and Microelectronics: Global Diffusion, Economics and Policy*, ed. N. Ekekwe (Hershey, Pennsylvania, US: Information Science Reference, IGI Global, 2010).

[7] Fuglie and Wang, "New Evidence Points to Robust but Uneven Productivity Growth in Global Agriculture"; Nin-Pratt, *Trends in Global Agriculture*.

[8] R. E. Evenson and D. Gollin, "Assessing the Impact of the Green Revolution: 1960–1980," *Science* 300, no. 5620 (2003): 758–762, doi:10.1126/science.1078710; R. E. Evenson and M. Rosegrant, "The Economic Consequences of Crop Genetic Improvement Programs," in *Crop Variety Improvement and Its Effect on Productivity: The Impact of International Agricultural Research*, ed. R. E. Evenson and D. Gollin (Wallingford, England: CAB International, 2003), doi:10.1079/9780851995496.0473; P. Pingali, "Will the Gene Revolution Reach the Poor? Lessons from the Green Revolution" (Mansholt

Lecture, Wageningen University, Germany, January 26, 2007); P. Pingali and P. W. Heisey, "Cereal-Crop Productivity in Developing Countries: Past Trends and Future Prospects," in *Agricultural Science Policy*, ed. J. M. Alston, P. G. Pardey, and M. Taylor (Washington, DC: IFPRI and Johns Hopkins University Press, 2001).

Cutting Consumer Food Waste

[1] J. Gustavsson, C. Cederberg, U. Sonesson, R. van Otterdijk, and A. Meybeck, *Global Food Losses and Food Waste: Extent, Causes and Prevention* (Rome: Food and Agriculture Organization of the United Nations, 2011).

[2] R. J. Hodges, J. C. Buzby, and B. Bennett, "Postharvest Losses and Waste in Developed and Developing Countries: Opportunities to Improve Resource Use," *Journal of Agricultural Sciences* 191, S1 (2010): 37–45.

[3] J. C. Buzby and J. Hyman, "Total and Per Capita Value of Food Loss in the United States," *Food Policy* 37 (2012): 561–571.

[4] Environmental Protection Agency, *Basic Information about Food Waste* (Washington, DC, 2012).

Reducing Postharvest Losses

[1] World Bank and FAO (Food and Agriculture Organization of the United Nations), *Missing Food: The Case of Postharvest Grain Losses in Sub-Saharan Africa* (Washington, DC: World Bank, 2011).

[2] J. Gustavsson, C. Cederberg, U. Sonesson, R. van Otterdijk, and A. Meybeck, *Global Food Losses and Food Waste: Extent, Causes and Prevention* (Rome: Food and Agriculture Organization of the United Nations, 2011).

What Makes African Agriculture Grow?

[1] A. Nin-Pratt, M. Johnson, and B. Yu, *Improved Performance of Agriculture in Africa South of the Sahara: Taking Off or Bouncing Back*, IFPRI Discussion Paper 1224 (Washington, DC: IFPRI, 2012).

[2] O. Badiane, *Sustaining and Accelerating Africa's Agricultural Growth Recovery in the Context of Changing Global Food Prices*, IFPRI Policy Brief 9 (Washington, DC: IFPRI, 2008).

Agricultural R&D: Spending Speeds Up

[1] N. Beintema, G-J. Stads, K. Fuglie, and P. Heisey, *ASTI Global Assessment of Agricultural R&D Spending: Developing Countries Accelerate Investment* (International Food Policy Research Institute, Washington, DC, 2012).

[2] All monetary values are measured in 2005 purchasing power parity (PPP) dollars.

[3] Please see "ASTI" in the "Food Policy Indicators" section of this book for details on definitions and country- and regional-level data.

CHAPTER 3

[1] World Commission on Environment and Development, *Our Common Future* (Oxford, UK: Oxford University Press, 1987), p. 43.

[2] Ibid., p. 46.

[3] United Nations Environment Programme, Agenda 21, 1992, www.unep.org/documents.multilingual/default.asp?documentid=52.

[4] F. Biermann, K. Abbott, S. Andresen, K. Bäckstrand, S. Bernstein, M. M. Betsill, H. Bulkeley, et al., "Navigating the Anthropocene: Improving Earth System Governance," *Science* 335, no. 6074 (2012): 1306–1307.

[5] United Nations Environment Programme, *Towards a Green Economy: Pathways to Sustainable Development and Poverty Eradication* (Nairobi, 2011), p. 16.

[6] World Bank, *Inclusive Green Growth: The Pathway to Sustainable Development* (Washington, DC, 2012), p. xi.

[7] Food and Agriculture Organization of the United Nations, *The State of Food Insecurity in the World 2012* (Rome, 2012), www.fao.org/docrep/016/i3027e/i3027e00.htm; World Bank, *Global Economic Prospects: Commodities at the Crossroads* (Washington, DC, 2009); M. W. Rosegrant, "Biofuels and Grain Prices: Impacts and Policy Responses," testimony for the US Senate Committee on Homeland Security and Governmental Affairs, May 7, 2008, www.ifpri.org/publication/biofuels-and-grain-prices; M. W. Rosegrant, T. Zhu, S. Msangi, and T. Sulser, "Global Scenarios for Biofuels: Impacts and Implications," *Review of Agricultural Economics* 30, no. 3 (2008): 495–505, http://dx.doi.org/10.1111/j.1467-9353.2008.00424.x.

[8] Millennium Ecosystem Assessment, *Ecosystems and Human Well-being: Synthesis* (Washington DC: Island Press, 2005).

[9] Food and Agriculture Organization of the United Nations, *State of the World's Forests* (Rome, 2011).

[10] Millennium Ecosystem Assessment, *Ecosystems and Human Well-being.*

[11] While total agricultural area increased by 10 percent during 1961–2009, per capita agricultural area declined by half, from 1.4 hectares to 0.7 hectare per person, over the same timeframe (Food and Agriculture Organization of the United Nations, FAOSTAT, Population and Resource Domains, http://faostat.fao.org, accessed October 29, 2012).

[12] E. Nkonya, N. Gerber, J. von Braun, and A. De Pinto, *Economics of Land Degradation: The Costs of Action versus Inaction*, IFPRI Issue Brief 68 (Washington, DC: International Food Policy Research Institute, 2011).

[13] W. Anseeuw, M. Boche, T. Breu, M. Giger, J. Lay, P. Messerli, and K. Nolte, *Transnational Land Deals for Agriculture in the Global South: Analytical Report Based on the Land Matrix Database* (Bern, Switzerland; Montpellier, France; and Hamburg, Germany: Centre for Development and Environment; CIRAD; and German Institute of Global and Area Studies, 2012).

[14] Millennium Ecosystem Assessment, *Ecosystems and Human Well-being: Current State and Trends: Findings of the Condition and Trends Working Group* (Washington, DC: Island Press, 2005); C. Ringler, "The Millennium Ecosystem Assessment: Tradeoffs between Food Security and the Environment," *Turkish Journal of Agriculture and Forestry* 32, no. 3 (2008): 147–157.

[15] M. W. Rosegrant, C. Ringler, and T. Zhu, "Water for Agriculture: Maintaining Food Security under Growing Scarcity," *Annual Review of Environment and Resources* 34 (2009): 205–223, doi: 10.1146/annurev.environ.030308.090351. For evidence on increased use of

micro-irrigation technologies, see International Commission on Irrigation and Drainage, *Annual Reports* (Delhi, various years).

[16] See Chapter 6 of this report.

[17] J. Huang, X. Wang, H. Zhi, Z. Huang, and S. Rozelle, "Subsidies and Distortions in China's Agriculture: Evidence from Producer-Level Data," *Australian Journal of Agricultural and Resource Economics* 55, no. 1 (2011): 53–71, www.ccap.org.cn/uploadfile/2011/0401/20110401065000946.pdf; M. Gopinath, "WTO Disciplines on Agricultural Support Update: A Comparison of India's WTO Domestic Support Notifications with Shadow Measurements" (Oregon State University, Corvallis, OR, US, 2012), mimeo.

[18] A. Bouët and D. Laborde Debucquet, *The Potential Cost of a Failed Doha Round*, IFPRI Discussion Paper 886 (Washington, DC: International Food Policy Research Institute, 2009); G. Nelson, A. Palazzo, C. Ringler, T. Sulser, and M. Batka, *The Role of International Trade in Climate Change Adaptation*, Issue Brief No. 4 (Geneva and Washington, DC: International Centre for Trade and Sustainable Development and International Food and Agricultural Trade Policy Council, 2009), www.agritrade.org/documents/IssueBrief4.pdf.

[19] CGIAR, *A Stripe Review of Natural Resources Management Research in the CGIAR* (Rome: CGIAR Independent Science and Partnership Council Secretariat, 2012).

[20] See, for example, B. Boxer, "Contradictions and Challenges in China's Water Policy Development," *Water International* 26, no. 3 (2001): 335–341.

[21] See, for example, N. Netzer, C. Detsch, A. Gross, H. Maihack, P. Oesterdiekhoff, and S. Wetzig, eds., *Really the Future We Want? Civil Society Voices on Rio+20* (Berlin: Friedrich Ebert Stiftung, 2012), http://library.fes.de/pdf-files/iez/global/09139.pdf; The Greens/European Free Alliance, The Future We REALLY Want, May 14, 2012, www.greens-efa.eu/the-future-we-really-want-7181.html.

[22] United Nations, Resolution 66/288, "The Future We Want," September 11, 2012, paragraph 56, http://sustainabledevelopment.un.org/futurewewant.html.

[23] Ibid., paragraph 111.

[24] Ibid., paragraph 66.

[25] As proposed by the governments of Colombia and Guatemala, Ministerio de Relaciones Exteriores, "Rio+20: Sustainable Development Goals: A Proposal from the Governments of Colombia and Guatemala," n.d., www.uncsd2012.org/rio20/content/documents/colombiasdgs.pdf.

[26] United Nations, "The Future We Want," paragraph 85.

[27] W. J. M. Tegart, G. W. Sheldon, and D. C. Griffiths, eds., *Climate Change: The IPCC Impacts Assessment* (Canberra: Australian Government Publishing Service, 1990); R. T. Watson, M. C. Zinyowera, and R. H. Moss, *Climate Change 1995: Impacts, Adaptations, and Mitigation of Climate Change*, IPCC Working Group I Report (Cambridge, UK: Cambridge University Press, 1996); M. L. Parry, O. F. Canziani, J. P. Palutikof, P. J. van der Linden, and C. E. Hanson, eds., *Climate Change 2007: Impacts, Adaptation, and Vulnerability*, Contribution of Working Group II to the Fourth Assessment Report of the Intergovernmental Panel on Climate Change (Cambridge, UK: Cambridge University Press, 2007).

[28] See, for example, World Business Council for Sustainable Development (WBCSD), *Vision 2050: The New Agenda for Business* (Geneva, 2010); and WBCSD, *Water Valuation: Building the Business Case* (Geneva, 2012).

[29] W. Klümper, J. Kathage, and M. Qaim, *Wahrnehmung des Themas Welternährung in der deutschen Öffentlichkeit*, Global Food Discussion Paper No. 16 (Göttingen, Germany: University of Göttingen, 2012), www.uni-goettingen.de/de/213486.html.

[30] United Nations, Agenda 21, paragraph 3.2.

[31] International Monetary Fund, *World Economic Outlook October 2012: Coping with High Debt and Sluggish Growth* (Washington DC, 2012), www.imf.org/external/pubs/ft/weo/2012/02/pdf/text.pdf.

Green and Greener: Toward Sustainable Agriculture

[1] FAO (Food and Agriculture Organization of the United Nations), *Declaration of the World Summit on Food Security* (Rome, 2009).

[2] UNEP (United Nations Environment Programme), *Towards a Green Economy: Pathways to Sustainable Development and Poverty Eradication* (Nairobi, 2011), www.unep.org/greeneconomy.

Ties That Bind Energy, Food, and Agriculture

[1] The energy focus is oil here because (1) it has a larger share than other energy sources; (2) it has multiple uses in transportation, electricity, and manufacturing; and (3) it is the reference price for other sources of energy. Actual prices used to inform these statements came from the International Monetary Fund.

CHAPTER 4

[1] Food and Agriculture Organization of the United Nations, *The State of Food and Agriculture 2010–2011: Women in Agriculture: Closing the Gender Gap for Development* (Rome, 2011).

[2] World Bank, World Development Report 2012 (Washington, DC, 2011).

[3] A. Peterman, A. R. Quisumbing, J. Behrman, and E. Nkonya, "Understanding the Complexities Surrounding Gender Differences in Agricultural Productivity in Nigeria and Uganda," *Journal of Development Studies* 47, no. 10 (2011): 1482–1509.

[4] C. Doss, C. D. Deere, A. D. Oduro, H. Swaminathan, J. Y. Suchitra, R. Lahoti, W. Baah-Boateng, L. Boakye-Yiadom, J. Contreras, J. Twyman, Z. Catanzarite, C. Grown, and M. Hillesland, *The Gender Asset and Wealth Gaps: Evidence from Ecuador, Ghana, and Karnataka, India* (Bangalore: Indian Institute of Management Bangalore, 2011).

[5] United States Agency for International Development, International Food Policy Research Institute, and Oxford Poverty and Human Development Initiative, Women's Empowerment in Agriculture Index, www.ifpri.org/publication/womens-empowerment-agriculture-index, accessed November 25, 2012.

[6] A. R. Quisumbing, B. Baulch, and N. Kumar, "Evaluating the Long-Term Impact of Antipoverty Interventions in Bangladesh: An Overview," *Journal of Development Effectiveness* 3, no. 2 (2011): 153–174.

[7] S. Lastarria-Cornhiel, J. A. Behrman, R. Meinzen-Dick, and A. R. Quisumbing, "Gender Equity and Land: Toward Secure and Effective Access for Rural Women," in *Gender in Agriculture and Food Security: Closing the Knowledge Gap*, edited by A. Quisumbing, R. Meinzen-Dick, T. Raney, A. Croppenstedt, J. A. Behrman, and A. Peterman (Dordrecht, Netherlands: Springer; and Rome: Food and Agriculture Organization of the United Nations, forthcoming).

[8] B. van Koppen, S. Smits, P. Moriarty, F. Penning de Vries, M. Mikhail, and E. Boelee, *Climbing the Water Ladder: Multiple-Use Water Services for Poverty Reduction* (The Hague, Netherlands: IRC International Water and Sanitation Centre and International Water Management Institute, 2009).

[9] R. Meinzen-Dick, A. Quisumbing, J. Behrman, P. Biermayr-Jenzano, V. Wilde, M. Noordeloos, C. Ragasa, and N. Beintema, *Engendering Agricultural Research*, IFPRI Research Monograph (Washington, DC: International Food Policy Research Institute, 2011).

[10] World Bank and International Food Policy Research Institute, *Gender and Governance in Rural Services: Insights from India, Ghana, and Ethiopia* (Washington, DC: 2010).

CHAPTER 5

[1] B. Losch, S. Fréguin-Gresh, and E. T. White, *Structural Transformation and Rural Change Revisited: Challenges for Late Developing Countries in a Globalizing World*, Africa Development Forum Series (Washington, DC: World Bank, 2012).

[2] The terms "young" and "youth" are used generally to include those aged 10 to 24 years old, depending on the context. Because different surveys and studies define them differently, it is difficult to establish one fixed group as "young."

[3] World Bank, *Agricultural Innovation Systems: An Investment Sourcebook* (Washington, DC: World Bank, 2012).

[4] "Young People, Farming and Food" conference, March 19–21, 2012, Accra, Ghana, www.future-agricultures.org/events/young-people-farming-a-food.

[5] African Union, "Report and Recommendations of the 4th Conference of the African Union Ministers in Charge of Youth Development," September 12–14, 2012, http://africa-youth.org/sites/default/files/FINAL%20REPORT%20COMY%20IV_ENGLISH.pdf.

[6] International Fund for Agricultural Development, "Youth in Agriculture: Special Session of the 2012 Farmers' Forum," February 18, 2012, www.ifad.org/farmer/2012/youth/index.htm.

[7] E. Crowley, "Employment-Centred Agricultural and Rural Development: Key to Reducing Poverty and Achieving Food Security," presentation for the United Nations 2012 Annual Ministerial Review, Global Preparatory Meeting, April 3, 2012, New York, www.un.org/en/ecosoc/newfunct/pdf/gpm_2012_crowley.pdf.

[8] Organisation for Economic Co-operation and Development, "Growing Prosperity: Agriculture, Economic Renewal, and Development," draft outcome document from the meeting "Innovating out of Poverty," Paris, April 6–7, 2009.

[9] World Bank, "Rural Transformation and Late Developing Countries in a Globalizing World: A Comparative Analysis of Rural Change," Economic and Sector Work Report, Agriculture and Rural Development Unit, Africa Region (Washington, DC, 2012).

[10] F. J. Proctor and V. Lucchesi, *Small-Scale Farming and Youth in an Era of Rapid Rural Change* (London and The Hague: International Institute for Environment and Development and Humanist Institute for Development Cooperation, 2012).

[11] For example, average maize yields in Africa are estimated to be 20 percent of the estimated potential yield. The rate is 32 percent for oil palm and soybeans and 54 percent for sugarcane; K. Deininger and D. Byerlee, "Rising Global Interest in Farm Land: Can It Yield Sustainable and Equitable Benefits?" (Washington, DC: World Bank, 2011). In other regions, the gap between actual and estimated yields for these crops seldom exceeds 40 percent.

[12] Losch, Fréguin-Gresh, and White, *Structural Transformation and Rural Change Revisited*.

[13] World Bank, "Agribusiness in Africa," draft AFTFP/AFTAR report (Washington, DC, 2012).

[14] N. Minot, *Transmission of World Food Price Changes to Markets in Sub-Saharan Africa*, IFPRI Discussion Paper No. 01059 (Washington, DC: International Food Policy Research Institute, 2011), www.ifpri.org/sites/default/files/publications/ifpridp01059.pdf.

[15] Organisation for Economic Co-operation and Development and Food and Agriculture Organization of the United Nations, *Agricultural Outlook 2011–2012* (Paris and Rome, 2012), Table B.1.

[16] O. Nagayets, "Small Farms: Current Status and Key Trends," information brief prepared for the "Future of Small Farms" workshop held at Wye College, June 26–29, 2005, Wye, UK, *The Future of Small Farms: Proceedings of a Research Workshop* (Washington, DC: International Food Policy Research Institute, 2005); J. Shaidi, "Youth Development in Tanzania," presentation, www.ilo.int/public/english/employment/recon/eiip/download/workshop/youthtan.pdf.

[17] World Bank internal staff analysis, 2012.

[18] A more complete treatment can be found in a forthcoming World Bank flagship report on employment in Africa.

[19] Proctor and Lucchesi, *Small-Scale Farming and Youth in an Era of Rapid Rural Change*.

[20] World Bank, "Transmigration in Indonesia," impact evaluation conducted by Independent Evaluation Group, accessed December 3, 2012, http://lnweb90.worldbank.org/oed/oeddoclib.nsf/DocUNIDViewForJavaSearch/4B8B0E01445D8351852567F5005D87B8.

[21] E. W. Chirwa, *Land Tenure, Farm Investments, and Food Production in Malawi*, Discussion Paper Series Number 18 (London: Department for International Development and Research Programme Consortium on Improving Institutions for Pro-Poor Growth, 2008).

[22] A. Nair, "Rural Leasing: An Alternative to Loans in Financing Income-Producing Assets," Brief 6 in *Innovations in Rural and Agriculture Finance*, edited by R. Kloeppinger-Todd and M. Sharma, 2020 Focus 18 (Washington, DC: International Food Policy Research Institute and World Bank, 2010).

[23] V. Mahajan and K. Vasumathi, "Combining Extension Services with Agricultural Credit: The Experience of BASIX India," Brief 13 in *Innovations in Rural and Agriculture Finance*, edited by R.

Kloeppinger-Todd and M. Sharma, 2020 Focus 18 (Washington, DC: International Food Policy Research Institute and World Bank, 2010).

[24] AgriFin, "Making Finance Work for Africa: Policy Brief on Agricultural Finance in Africa" (Addis Ababa and Berlin: African Union, Germany Agency for International Cooperation [GIZ], and German Federal Ministry for Economic Cooperation [BMZ], 2012); R. Kloeppinger-Todd and M. Sharma, ed., *Innovations in Rural and Agriculture Finance*, 2020 Focus 18 (Washington, DC: International Food Policy Research Institute and World Bank, 2010).

[25] G. van Empel, "Rural Banking in Africa: The Rabobank Approach," Brief 4 in *Innovations in Rural and Agriculture Finance*, ed. R. Kloeppinger-Todd and M. Sharma, 2020 Focus 18 (Washington, DC: International Food Policy Research Institute and World Bank, 2010).

[26] J. Campaigne and T. Rausch, "Bundling Development Services with Agricultural Finance: The Experience of DrumNet," Brief 14 in *Innovations in Rural and Agriculture Finance*, ed. R. Kloeppinger-Todd and M. Sharma, 2020 Focus 18 (Washington, DC: International Food Policy Research Institute and World Bank, 2010).

[27] Rabobank Group, Rabo Sustainable Agriculture Guarantee Fund, accessed December 3, 2012, www.rabobank.com/content/products_services/business_clients/professionalproducts/raboagrifund/index.jsp.

[28] AgriFin, "Making Finance Work for Africa."

[29] S. Lonie, "M-PESA: Finding New Ways to Serve the Unbanked in Kenya," Brief 8 in *Innovations in Rural and Agriculture Finance*, ed. R. Kloeppinger-Todd and M. Sharma, 2020 Focus 18 (Washington, DC: International Food Policy Research Institute and World Bank, 2010).

[30] X. Giné, "Biometric Technology in Rural Credit Markets: The Case of Malawi," Brief 9 in *Innovations in Rural and Agriculture Finance*, ed. R. Kloeppinger-Todd and M. Sharma, 2020 Focus 18 (Washington, DC: International Food Policy Research Institute and World Bank, 2010).

[31] World Bank, "Land Administration and Reform in Sub-Saharan Africa: From Piloting to Scaling Up," Economic Sector Work Report (Washington, DC, 2012).

[32] Ibid.

[33] Ibid.

[34] K. Deininger, *Land Policies for Growth and Poverty Reduction* (Washington, DC: World Bank, 2003).

[35] World Bank, "Land Administration and Reform in Sub-Saharan Africa."

[36] E. Lahiff and G. Li, "Land Redistribution in South Africa: A Critical Review," working paper for the "Land Administration and Reform in SSA" study (World Bank, Washington, DC, 2012).

[37] H. Tchale, "Pilot Redistributive Land Reform in Malawi: Innovations and Emerging Good Practices," working paper for the "Land Administration and Reform in SSA" study (World Bank, Washington, DC, 2012).

[38] Kenya, National Agricultural Sector Extension Policy (NASEP) (Nairobi: Ministry of Agriculture, Ministry of Livestock and Fisheries Development, and Ministry of Cooperative Development and Marketing, 2005).

[39] World Bank, *Agricultural Innovation Systems: An Investment Sourcebook* (Washington, DC: World Bank, 2012).

[40] Comprehensive Africa Agriculture Development Programmme, www.nepad-caadp.net/.

[41] Proctor and Lucchesi, *Small-Scale Farming and Youth in an Era of Rapid Rural Change*.

India's New Deal: Public Works and Rural Jobs

[1] India, Ministry of Rural Development, *MGNREGA Sameeksha: An Anthology of Research Studies on Mahatma Gandhi National Rural Employment Guarantee Act 2005, 2006–2012* (Orient BlackSwan: New Delhi, 2012).

[2] P. Kumar and P. K. Joshi, *Progress Report on MGNREGA and Household Nutritional Security: Trade, Agricultural Policies and Structural Changes in India's Agri-food System: Implications for National and Global Markets* (International Food Policy Research Institute: Washington, DC, forthcoming).

CHAPTER 6

[1] US Constitution, article I, section 9.

[2] R. M. Chite et al., *The 2012 Farm Bill: A Comparison of Senate-Passed S-3240 and the House Agriculture Committee's H.R. 6083 with Current Law*, CRS Report for Congress R42552 (Washington, DC: Congressional Research Service, 2012).

[3] Zulauf and Orden argue that US crop insurance subsidies exceed levels justified on systemic risk grounds by comparing subsidy rates to the degree of correlation of individual farm revenue risk to revenue risks at the county, state, and national levels. See C. Zulauf and D. Orden, *US Farm Policy and Risk Assistance: The Competing Senate and House Agriculture Committee Bills of July 2012*, Issue Paper No. 44 (Geneva: International Centre for Trade and Sustainable Development, 2012), http://ictsd.org/i/publications/144525/. Other critical analysts question any public rationale for crop insurance subsidies; see V. H. Smith, *Premium Payments: Why Crop Insurance Costs Too Much* (Washington, DC: American Enterprise Institute, 2011), www.aei.org/files/2011/11/04/-premium-payments-why-crop-insurance-costs-toomuch_152221377467.pdf.

[4] Zulauf and Orden, *US Farm Policy and Risk Assistance*.

[5] World Trade Organization, Agreement on Agriculture, Annex 2, www.wto.org/english/docs_e/legal_e/14-ag_02_e.htm#annII.

[6] D. Orden, D. Blandford, and T. Josling, *WTO Disciplines on Agricultural Support: Seeking a Fair Basis for Trade* (Cambridge, UK: Cambridge University Press, 2011).

[7] J.-P. Butault, J.-C. Bureau, H.-P. Witzke, and T. Heckelei, *Comparative Analysis of Agricultural Support within the Major Agricultural Trading Nations* (Brussels: European Parliament, 2012), www.europarl.europa.eu/committees/fr/studiesdownload.html?languageDocument=EN&file=74651.

[8] A. Matthews, *Environmental Public Goods in the New CAP: Impact of Greening Proposals and Possible Alternatives* (Brussels: European Parliament, 2012), www.europarl.europa.eu/committees/fr/studiesdownload.html?languageDocument=EN&file=74995.

[9] L.-P. Mahé, *Do the Proposals for the CAP after 2013 Herald a "Major" Reform?* Policy Paper 53 (Paris: Notre Europe, 2012).

CHAPTER 7

[1] AGIR (Global Alliance for Resilience in the Sahel), *Feuille de Route (Version provisoire)*, www.oecd.org/fr/csao/dossiers/FR-AGIR draft roadmap_191112.pdf, accessed January 15, 2013.

[2] Economic Community of West African States, *Réserve Régionale de Sécurité Alimentaire* (Abuja, Nigeria, 2012).

[3] O. Badiane, S. Odjo, F. Wouterse, and J. Ulimwengu, "Comparative Analysis of Strategies and Long Term Outlook for Growth and Poverty Reduction among ECOWAS Member Countries," proceedings presented at the Theory, Methods and Application of Modeling and Evaluation (M&E) of Comprehensive Africa Agriculture Development Programme (CAADP) Policies Workshop, Kiel, Germany, July 2–3, 2011.

[4] S. Benin, A. Nin Pratt, S. Wood, and Z. Guo, *Trends and Spatial Patterns in Agricultural Productivity in Africa: 1961–2010*, ReSAKSS Annual Trends and Outlook Report 2011 (Washington DC: International Food Policy Research Institute, 2012).

[5] Global Agriculture and Food Security Program, 2012 Annual Report, www.gafspfund.org/gafsp/sites/gafspfund.org/files/Documents/GAFSP_AR12_final.pdf, accessed January 14, 2013.

[6] A. Dorward and E. Chirwa, "The Malawi Agricultural Input Subsidy Programme: 2005/06 to 2008/09," *International Journal of Agricultural Sustainability* 16 (2011): 232–247.

[7] World Bank, *Middle East and North Africa Region Economic Developments and Prospects, October 2012: Looking Ahead after a Year in Transition* (Washington, DC: The World Bank, 2012).

[8] Food and Agriculture Organization of the United Nations, Global Information and Early Warning System, "Egypt," www.fao.org/giews/countrybrief/country.jsp?code=EGY, accessed January 14, 2013; Food and Agriculture Organization of the United Nations, Global Information and Early Warning System, "Libya," www.fao.org/giews/countrybrief/country.jsp?code=LBY, accessed January 14, 2013; Food and Agriculture Organization of the United Nations, Global Information and Early Warning System, "Tunisia," www.fao.org/giews/countrybrief/country.jsp?code=TUN, accessed January 14, 2013; Food and Agriculture Organization of the United Nations, Global Information and Early Warning System, "Yemen," www.fao.org/giews/countrybrief/country.jsp?code=YEM, accessed January 14, 2013.

[9] Food and Agriculture Organization of the United Nations, The State of Food Insecurity in the World, www.fao.org/publications/sofi/en/, accessed January 22, 2013. Western Asia includes Iraq, Jordan, Kuwait, Lebanon, Saudi Arabia, Syria, Turkey, United Arab Emirates, and Yemen. The number of undernourished in these countries increased from 18 million to 21 million between 1990–1992 and 2010–2012. North Africa includes Algeria, Egypt, Libya, Morocco, and Tunisia. The number of undernourished in these countries stayed at 4 million in the observed time periods.

[10] The government of Jordan raised fuel prices in September of 2012. However, in the face of significant protests, King Abdullah II ordered a partial freeze on these price hikes. See Economist Intelligence Unit, Jordan Country Report, https://portal.eiu.com/login.aspx?service=https://www.eiu.com&brand=&clientid=44bs26t12v11t&gateway=true&returnto=https%3a%2f%2fwww.eiu.com%2findex.asp%3f%26, accessed October 2012.

[11] I. Coleman, "Reforming Petroleum Subsidies in Yemen and Egypt," *Democracy in Development* (blog), February 8, 2012, http://blogs.cfr.org/coleman/2012/02/08/reforming-petroleum-subsidies-in-yemen-and-egypt/, accessed January 22, 2013; International Monetary Fund, *Arab Countries in Transition: Economic Outlook and Key Challenges*, Deauville Partnership Ministerial Meeting, October 12, 2012, Tokyo (Washington, DC: International Monetary Fund), www.imf.org/external/region/mcd/deauville/note101212.pdf, accessed January 14, 2013.

[12] World Food Programme and the Egyptian Cabinet Information and Decision Support Center, "Egyptian Food Observatory: Food Monitoring and Evaluation System, *Quarterly Bulletin*, Issue 8: April–June 2012, http://documents.wfp.org/stellent/groups/public/documents/ena/wfp251436.pdf, accessed January 14, 2013. Baladi bread prices have been fixed since 1989. For rice, sugar, oil, and tea, Egypt currently has a food subsidy system for which the government in 2011 issued 15.6 million ration cards to eligible households, covering close to 60 million individuals at a cost of £E0.25 per loaf.

[13] Food and Agriculture Organization of the United Nations, Committee on Food Security, "39th Session, Rome, Italy, 15–20 October 2012," www.fao.org/bodies/cfs/cfs39/en/, accessed January 14, 2013.

[14] Y. B. Thapa, *Constraints and Approach for Improving Fertilizer Supply for Meeting Domestic Demand* (Kathmandu, Nepal: Economic Policy Network, Government of Nepal, 2006).

[15] T. Reardon and B. Minten, *The Quiet Revolution in India's Food Supply Chain*, IFPRI Discussion Paper 01115 (Washington DC: International Food Policy Research Institute, 2011).

[16] P. K. Joshi, A. Gulati, and R. Cummings Jr., "Agricultural Diversification in South Asia: Beyond Food Security," in *Agricultural Diversification and Smallholders in South Asia*, ed. P. K. Joshi, A. Gulati, and R. Cummings Jr (New Delhi, India: Academic Foundation, 2007).

[17] A. V. Jakhar, "FDI in Multi Brand Retail: Still No Center Stage for the Farmers," *Farmers' Forum* 12, no. 5 (September–October 2012): 1–9.

[18] K. Chen and P. K. Joshi, "New Players: Stepping into the Global Food System," in *2011 Global Food Policy Report* (Washington DC: International Food Policy Research Institute, 2011).

[19] A. Burns and T. Janse van Rensburg, *Global Economic Prospects*, Vol. 5 (Washington, DC: World Bank, 2012).

[20] Asian Development Bank, *How Can Asia Respond to Global Economic Crisis and Transformation?* (Mandaluyong City, Philippines: 2012).

[21] Asian Development Bank, *Asian Development Outlook 2012: Confronting Rising Inequality in Asia* (Mandaluyong City, Philippines: 2012).

[22] This is based on a 2012 rice-mill survey's findings in the provinces of Jiang and Zhejiang, China. The survey project is jointly supported by the Asian Development Bank and International Food Policy Research Institute, ADB-IFPRI Project RETA 14 (7648).

[23] Author's calculations based on US Department of Agriculture Economic Research Service, "U.S. Drought 2012: Farm and Food Impacts,"

www.ers.usda.gov/topics/in-the-news/us-drought-2012-farm-and-food-impacts.aspx, accessed November 21, 2012.

[24] E. Samboh, "Govt Scraps Import Duties on Soybeans to Stabilize Soaring Prices," *The Jakarta Post*, July 26, 2012, www.thejakartapost.com/news/2012/07/26/govt-scraps-import-duties-soybeans-stabilize-soaring-prices.html, accessed November 21, 2012; *The Jakarta Post*, "Economy in Brief: Bulog to Manage 500,000 Tons of Soybeans," October 10, 2012, www.thejakartapost.com/news/2012/10/10/economy-brief-bulog-manage-500000-tons-soybeans.html, accessed on November 21, 2012.

[25] US Department of Agriculture Foreign Agricultural Service, *Grain: World Markets and Trade*, Circular Series November Issue (Washington, DC: 2012).

[26] K. Z. Chen, K. Flaherty, and Y. Zhang, *China: Recent Developments in Agricultural Research*, Agricultural Science and Technology Indicators Country Note (Washington, DC: International Food Policy Research Institute, 2012); Agricultural Science and Technology Indicators, "China Accelerates Agricultural Technology Development," http://astinews.ifpri.info/2012/08/21/china-accelerates-agricultural-technology-development/, accessed November 21, 2012.

[27] J. Szep, "Myanmar Seeks to Rejoin Top Rice Exporters," Reuters, August 9, 2012, http://in.reuters.com/article/2012/08/09/myanmar-rice-idINL4E8IJ21O20120809, accessed November 21, 2012.

[28] US Department of Agriculture, *Grain: World Markets and Trade*, Foreign Agricultural Service Circular Series FG 01-13 (Washington, DC, 2013).

[29] ASEAN, "The Twelfth Meeting of The ASEAN Ministers of Agriculture and Forestry and the Ministers of Agriculture of The People's Republic of China, Japan and the Republic of Korea (12th AMAF Plus Three)," joint press statement, September 28, 2012, www.asean.org/news/asean-secretariat-news/item/the-twelfth-meeting-of-the-asean-ministers-of-agriculture-and-forestry-and-the-ministers-of-agriculture-of-the-people-s-republic-of-china-japan-and-the-republic-of-korea-12th-amaf-plus-three, accessed January 2, 2013.

[30] ASEAN, Overview ASEAN Plus Three Cooperation, www.asean.org/news/item/asean-plus-three-cooperation, accessed November 29, 2012.

[31] APTERR, History, www.apterr.org/index.php/history, accessed November 29, 2012; APTERR, How APTERR Works? http://www.apterr.org/index.php/how-apterr-works, accessed November 29, 2012.

[32] R. M. Briones, A. Durand-Morat, E. J. Wailes, and E. C. Chavez, *Climate Change and Price Volatility: Can We Count on the ASEAN plus Three Emergency Rice Reserve?* ADB Sustainable Development Working Paper Series (Mandaluyong City, Philippines: Asian Development Bank, 2012).

[33] Comisión Económica para América Latina y el Caribe, Food and Agriculture Organization of the United Nations, Instituto Interamericano de Cooperación para la Agricultura, Perspectivas de la Agricultura y del Desarrollo Rural en las Américas: Una Mirada Hacia América Latina y el Caribe, www.eclac.org/publicaciones/xml/9/48259/perspectivas-2013.pdf, accessed January 14, 2013.

[34] Reuters, "Brazil GMO Planting to Increase by 14 Percent This Season," December 17, 2012, www.reuters.com/article/2012/12/17/brazil-biotech-idUKL1E8NH1LW20121217, accessed January 10, 2013.

[35] Comisión Económica para América Latina y el Caribe, Food and Agriculture Organization of the United Nations, Instituto Interamericano de Cooperación para la Agricultura, Perspectivas de la Agricultura y del Desarrollo Rural en las Américas.

Vision 2021: Bangladesh Charts a Path toward Food Security

[1] Bangladesh Bureau of Statistics, Household Income and Expenditure Survey (Dhaka, 2010), www.bbs.gov.bd/PageWebMenuContent.aspx?MenuKey=320.

CHAPTER 8

[1] P. Abbott, C. Hurt, and W. Tyner, "What's Driving Food Prices in 2011?" Farm Foundation Issue Report (July 2011), www.farmfoundation.org/news/articlefiles/1742-FoodPrices_web.pdf; D. Headey, S. Malaiyandi, and S. Fan, *Navigating the Perfect Storm: Reflections on the Food, Energy, and Financial Crisis*, IFPRI Discussion Paper 889 (Washington, DC: International Food Policy Research Institute, 2009); A. Mittal, "The Blame Game," in *The Global Food Crisis*, ed. J. Clapp and M. Cohen (Waterloo, Ontario, Canada: Centre for International Governance Innovation and Wilfrid Laurier University Press, 2009); T. Yu, S. Tokgoz, E. Wailes, and E. Chavez, "A Quantitative Analysis of Trade Policy Responses to Higher World Agricultural Commodity Prices," Food Policy 36, no. 5 (2011): 545–561.

[2] World Bank, GEM Commodities, http://data.worldbank.org/data-catalog/commodity-price-data, accessed January 18, 2013.

[3] H. Westhoek, T. Rood, M. van den Berg, J. Janse, D. Nijdam, M. Reudink, and E. Stehfest, *The Protein Puzzle: The Consumption and Production of Meat, Dairy, and Fish in the European Union* (The Hague: Netherlands Environmental Assessment Agency, 2011).

[4] IMPACT covers more than 46 crops and livestock commodities. It includes 115 countries, with each country linked to the rest of the world through international trade, and 281 food-producing units (grouped according to political boundaries and major river basins). Demand is a function of prices, income, and population growth. Population projections are the "medium" variant population growth rate projections from the Population Division of the United Nations. Gross domestic product (GDP) projections are estimated by the authors, drawing upon Millennium Ecosystem Assessment, *Ecosystems and Human Well-Being: General Synthesis*, Millennium Ecosystem Assessment Series (Washington, DC: Island Press, 2005). Crop production is determined by crop and input prices, the rate of productivity growth, and water availability. For more details on the model, see M. W. Rosegrant and the IMPACT Development Team, "International Model for Policy Analysis of Agricultural Commodities and Trade (IMPACT): Model Description" (Washington, DC: International Food Policy Research Institute, 2012), www.ifpri.org/sites/default/files/publications/impactwater2012.pdf.

[5] S. M. Borras, "Questioning Market-Led Agrarian Reform: Experiences from Brazil, Colombia, and South Africa," *Journal of Agrarian*

Change 3, no. 3 (2003): 367–394; N. Beintema, A. Flavio Dias Avila, and P. G. Pardey, Agricultural R&D in Brazil: Policy, Investments, and Institutional Profile (Washington, DC: IFPRI, Embrapa, and FONTAGRO, 2001), http://www.ifpri.org/themes/grp01/grp01_brazil.pdf, accessed July 14, 2006; J. Brooks and O. Melyukhina, "The Effects of Agricultural Policy Reform on Poverty in Brazil," Workshop on Agricultural Policy Reform and Adjustment, Imperial College, Wye, UK, October 23–25, 2003; E. Dohlman, S. Osborne, and B. Lohmar, "Dynamics of Agricultural Competitiveness: Policy Lessons from Abroad," *Amber Waves* (Economic Research Service, USDA) 1, no. 2 (2001): 14–21; S. Helfand and G. Castro de Rezende, "Brazilian Agriculture in the 1990s: Impact of the Policy Reforms," UC Riverside Economics Working Paper No. 01-34, 2001; US Department of Agriculture, Economic Research Service, "Brazil: Policy" (May 30, 2012), www.ers.usda.gov/topics/international-markets-trade/countries-regions/brazil/policy.aspx#research, accessed January 3, 2013; US Department of Agriculture, Foreign Agricultural Service, *Record Brazilian Agricultural Production Spurs Further Export Gains*, International Agricultural Trade Report (Washington, DC, 2012), www.fas.usda.gov/info/IATR/012412_Brazil/012412_Brazil.pdf; World Food Prize Foundation, "2006 World Food Prize Winners Opened Brazil's 'Closed Lands,'" press release, June 15, 2006, www.worldfoodprize.org/index.cfm?nodeID=24667&audienceID=1&action=display&newsID=8086; S. Zahniser, D. Pick, G. Pompelli, and M. Gehlhar, "Trade Liberalization in the Western Hemisphere: Impacts on US Agricultural Exports," in *US Agriculture and the Free Trade Area of the Americas*, ed. M. E. Burfisher, Agricultural Economic Report No. 827 (Washington, DC: US Department of Agriculture, Economic Research Service, 2004).

[6] Food and Agriculture Organization of the United Nations, FAOSTAT, 2012.

[7] Rosegrant et al., "IMPACT: Model Description." To estimate income projections, we drew upon scenario drivers from the Millennium Ecosystem Assessment (see note 4) and updated them based on recent data on economic performance. We used population projections from the "medium" variant projections of population growth from the Population Division of the United Nations.

[8] FAO, FAOSTAT, 2012.

[9] The percentage of malnourished children younger than the age of five is estimated from average per capita calorie consumption, female access to secondary education, quality of maternal and child care, and health and sanitation; Rosegrant et al., "IMPACT: Model Description." The relationship used to project the percentage of malnourished children is based on a cross-country regression relationship from L. Smith and L. Haddad, *Explaining Child Malnutrition in Developing Countries: A Cross-Country Analysis*, IFPRI Research Report 111 (Washington, DC: International Food Policy Research Institute, 2000). The data used to make this calculation are obtained from the World Health Organization's Global Database on Child Growth Malnutrition, the United Nations Administrative Committee on Coordination–Subcommittee on Nutrition, the World Bank's World Development Indicators, the FAOSTAT database, and the UNESCOSTAT database. The population at risk of hunger is computed using the share of the population at risk of hunger, which is based on a strong empirical correlation between the share of malnourished people within the total population and the relative availability of food. It is adapted from the work in G. Fischer, M. Shah, F. N. Tubiello, and H. van Velhuizen, "Socio-economic and Climate Change Impacts on Agriculture: An Integrated Assessment," *Philosophical Transactions of the Royal Society B* 360, no. 1463 (2005): 2067–2083.

[10] J. Alston, T. J. Wyatt, P. G. Pardey, M. C. Marra, and C. Chan-Kang, *A Meta-Analysis of Rates of Return to Agricultural R&D*, IFPRI Research Report 113 (Washington, DC: International Food Policy Research Institute, 2002).

[11] B. M. Popkin, "Nutritional Patterns and Transitions," *Population and Development Review* 19, no. 1 (1993): 138–157; A. Drewnowski and B. M. Popkin, "The Nutrition Transition: New Trends in the Global Diet," *Nutrition Reviews* 55, no. 2 (1997): 31–43.

[12] J. Huang and H. Bouis, *Structural Changes in the Demand for Food in Asia*, 2020 Vision Brief 41 (Washington, DC: International Food Policy Research Institute, 1996); World Health Organization, *Diet, Nutrition, and the Prevention of Chronic Diseases: Report of a Joint WHO/FAO Consultation*, WHO Technical Report Series 916 (Geneva, 2003).

[13] C. Delgado, M. Rosegrant, H. Steinfeld, S. Ehui, and C. Courbois, *Livestock to 2020: The Next Food Revolution*, 2020 Discussion Paper 28 (Washington, DC: International Food Policy Research Institute, jointly with the Food and Agriculture Organization of the United Nations and the International Livestock Research Institute, 1993).

[14] J. Bruinsma, ed., *World Agriculture towards 2015/2030: An FAO Perspective* (Rome: Food and Agriculture Organization of the United Nations; London: Earthscan, 2003).

[15] M. McCullough, D. Feskanich, M. J. Stampfer, E. L. Giovannucci, E. B. Rimm, F. B. Hu, D. Spiegelman, D. J. Hunter, G. A. Colditz, and W. C. Willett, "Diet Quality and Major Chronic Disease Risk in Men and Women: Moving toward Improved Dietary Guidance," *American Journal of Clinical Nutrition* 76, no. 2 (2002): 1261–1271.

[16] H. Steinfeld, P. Gerber, T. Wassenaar, V. Castel, M. Rosales, and C. de Haan, *Livestock's Long Shadow: Environmental Issues and Options* (Rome: Food and Agriculture Organization of the United Nations, 2006).

[17] See, for example, J. A. Foley, R. Navin, K. A. Brauman, E. S. Cassidy, J. S. Gerber, M. Johnston, N. D. Mueller, C. O'Connell, D. K. Ray, P. C. West, C. Balzer, E. M. Bennett, S. R. Carpenter, J. Hill, C. Monfreda, S. Polasky, J. Rockstrom, J. Sheehan, S. Siebert, D. Tilman, and D. P. M. Zaks, "Solutions for a Cultivated Planet," *Nature* 478, no. 7569 (2011): 337–342; M. Gold, *The Global Benefits of Eating Less Meat: A Report for Compassion in World Farming Trust* (Surrey, UK: Compassion in World Farming Trust, 2004), www.ciwf.org.uk/includes/documents/cm_docs/2008/g/global_benefits_of_eating_less_meat.pdf.

[18] The IMPACT model infers a trend in levels of malnutrition among the most vulnerable demographic of the population: children under five. The determinants of malnutrition are derived primarily from four key indicators, which were first established by Smith and Haddad in their cross-country estimation work: per capita calorie availability, access to clean drinking water, rates of secondary schooling among females, and the ratio of female-to-male life expectancy. These determinants are consistent with the four-pillared concept of food security underlying FAO's conceptual framework, where availability is only one factor accounting for food security status among vulnerable populations, and must be evaluated along with access,

utilization, and stability. The methodology used for tracking child malnutrition in IMPACT is based on this work and is implemented through an analytical relationship that is parameterized by the statistical coefficients derived by Smith and Haddad's work. Smith and Haddad, *Explaining Child Malnutrition in Developing Countries*; Food and Agriculture Organization of the United Nations, "Declaration of the World Summit on Food Security," Rome, November 16–18, 2009; O. Ecker and C. Breisinger, *The Food Security System: A New Conceptual Framework,* IFPRI Discussion Paper 1166 (Washington, DC: International Food Policy Research Institute, 2012).

[19] The IMPACT model does not disaggregate demand to various substrata of the population and thus does not show explicitly what the reduction in meat consumption in Brazil and China would mean for the distributional intake of meat within the population. However, given that the wealthier sections of the population in these countries eat much more meat per capita than the poorer members of the population, their reduction in meat consumption growth means that the poorer households might actually be able to consume more. In other words, the reduction in the price of meat products made possible by lowered consumption among the wealthier members of society would make meat more affordable for others. So even if poorer households slightly increase their meat intake, an overall national reduction of per capita meat intake could occur through the redistribution of consumption within the population.

Modeling the Future: How Can We Improve Food Policy?

[1] For more information on the CGIAR Research Program on Policies, Institutions and Markets, visit www.pim.cgiar.org.

[2] For more information, visit www.ifpri.org/pressrelease/global-futures.

FOOD POLICY INDICATORS: TRACKING CHANGE

[1] International Food Policy Research Institute, *2011 Global Food Policy Report* (Washington, DC, 2012).

[2] For details on data used see K. O. Fuglie, "Productivity Growth and Technology Capital in the Global Agricultural Economy," in *Productivity Growth in Agriculture: An International Perspective*, ed. K. Fuglie, S. L. Wang, and V. E. Ball (Wallingford, Oxon, UK: CAB International, 2012).

[3] J. Howard, "A Push Grows Abroad for Open Access to Publicly Financed Research," *Chronicle of Higher Education*, August 13, 2012.

New Food Security Indexes

[1] DuPont, "DuPont Calls for Common Food Security Metrics," news release, July 10, 2012, http://foodsecurity.dupont.com/2012/07/10/dupont-calls-for-common-food-security-metrics/; Syngenta, "Rice Bowl Index Highlights Solutions for Food Security Challenges across Asia-Pacific," news release, May 30, 2012, www.syngenta.com/global/corporate/en/news-center/news-releases/Pages/120530.aspx, accessed December 28. 2012; Danish International Development Agency (Danida)/US Agency for International Development, "Consultation Report," Copenhagen Consultation, Denmark, June 28–29, 2012, www.agriculturaltransformationindex.org, accessed December 28, 2012; M. Garde, K. Eardley, L. Holly, and L. O'Shea, *The Nutrition Barometer* (London: Save the Children, 2012), http://everyone.org/nutrition-barometer, accessed December 28, 2012.

[2] Institute of Development Studies, Hunger Reduction Commitment Index, www.hrcindex.org/category/background-documents/, accessed December 28, 2012; Maplecroft, "Differing Fortunes of Key Economies Revealed by Maplecroft's Food Security Index," news release, September 3, 2009, http://maplecroft.com/about/news/food_security_research_reveals_differing_fortunes_of_key_economies_04.html, accessed December 28, 2012.

[3] Food and Agriculture Organization of the United Nations, *The State of Food Insecurity in the World 2012* (Rome: 2012), www.fao.org/docrep/016/i3027e/i3027e00.htm, accessed December 28, 2012; US Department of Agriculture (USDA), GFA: Food Security Assessment Situation and Outlook Catalog, www.ers.usda.gov/publications/gfa-food-security-assessment-situation-and-outlook, accessed September 11, 2012; World Health Organization (WHO), Global Burden of Disease, www.who.int/topics/global_burden_of_disease, accessed September 11, 2012.

[4] International Food Policy Research Institute, Africa Food Security Vulnerability Indices, www.foodsecurityportal.org/africa-food-security-vulnerability-indices, accessed September 11, 2012; International Food Policy Research Institute, Global Hunger Index, www.ifpri.org/ourwork/researcharea/global-hunger-index, accessed September 11, 2012.

[5] D. Headey and O. Ecker, *Improving the Measurement of Food Security*, IFPRI Discussion Paper 1225 (Washington, DC: International Food Policy Research Institute, 2012), www.ifpri.org/publication/improving-measurement-food-security, accessed January 9, 2013.

[6] International Food Policy Research Institute, Women's Empowerment in Agriculture Index, www.ifpri.org/publication/womens-empowerment-agriculture-index, accessed November 2, 2012; Agricultural Science and Technology Indicators, www.asti.cgiar.org, accessed September 11, 2012; World Bank, Living Standards Measurement Study, http://go.worldbank.org/IPLXWMCNJ0, accessed September 18, 2012.

[7] Food and Agriculture Organization of the United Nations, *State of Food Insecurity in the World 2012.*

Knowledge Is Power: Open Access in 2012

[1] T. Brody, *Registry of Open Access Repositories Mandatory Archiving Policies (ROARMAP)*, (Southampton, England: University of Southampton, School of Electronics and Computer Science, 2012), http://roarmap.eprints.org, accessed January 25, 2013.

[2] K. Shearer, *Confederation of Open Access Repositories (COAR)* (Göttingen, Germany: Göttingen State and University Library, 2009), www.coar-repositories.org, accessed January 25, 2013.

[3] J. Howard, "A Push Grows Abroad for Open Access to Publicly Financed Research," *Chronicle of Higher Education*, August 13, 2012.

[4] European Commission, "Scientific Data: Open Access to Research Results Will Boost Europe's Innovation Capacity," press release, July 17, 2012.

[5] India, Ministry of Science & Technology, 2013, "National Data Sharing and Accessibility Policy (NSDAP)," www.dst.gov.in/nsdi.html, accessed January 24.

[6] UK Department for International Development, *DFID Research Open and Enhanced Access Policy* (London: 2013), www.dfid.gov.uk/Documents/publications1/DFIDResearch-Open-and-Enhanced-Access-Policy.pdf.

[7] Research Councils UK, "Research Councils UK Policy on Access to Research Outputs," www.rcuk.ac.uk/documents/documents/RCUK%20_Policy_on_Access_to_Research_Outputs.pdf, accessed January 24, 2013.

[8] Wellcome Trust, "Wellcome Trust Strengthens Its Open Access Policy," press release, June 28, 2012.

[9] White House, Petition to the Obama Administration, "Require Free Access over the Internet to Scientific Journal Articles Arising from Taxpayer-Funded Research," May 13, 2012, https://petitions.whitehouse.gov/petition/require-free-access-over-internet-scientific-journal-articles-arising-taxpayer-funded-research/wDX82FLQ?utm_source=wh.gov&utm_medium=shorturl&utm_campaign=shorturlb.

[10] World Bank, "World Bank Announces Open Access Policy for Research and Knowledge, Launches Open Knowledge Repository," press release, April 10, 2012, http://web.worldbank.org/WBSITE/EXTERNAL/NEWS/0,,contentMDK:23164491~pagePK:64257043~piPK:437376~theSitePK:4607,00.html.

[11] International Food Policy Research Institute, "Open Access Statement," www.ifpri.org/ourwork/about/open-access-statement, accessed February, 8, 2013.